D0126134

"Sherrell is a passionate advocate veys with urgency and honest, ra has infiltrated the essence of his being. He writes with a frightening sense of gravity. . . . *Warmth* is a pleading, informative call to action."

—*BookPage*

"Insightful reflections from a thoughtful, energetic activist."

—*Kirkus Reviews*

"Sherrell's strikingly perceptive book is neither a prescription for hope nor for despair, but a call for a clear-eyed examination of one of the most pressing questions of our time—what do we owe the next generation?"

—Jenny Offill, *New York Times* bestselling author of *Weather* and *Dept. of Speculation*

"Beautifully rendered and bracingly honest, this book helped me do the impossible: live in the space between grief and hope."

—Jenny Odell, *New York Times* bestselling author of *How to Do Nothing*

"Searchingly honest, this fine book is the work of someone actively engaged in the most important fight of our time (maybe of all time), and also of a writer able to establish the necessary distance. Dan Sherrell is smart, obviously, but he's also something much more important: open, vulnerable, able to face fully that which we all must grapple with in this overheating century."

—Bill McKibben, *New York Times* bestselling author of *Falter*

"In this insider account of the struggle for the earth against the forces of corporate greed that threaten it, Daniel Sherrell has written a tender letter to the uncertain future—at once intimate and angry, exasperated and brave."   —Anne Boyer, Pulitzer Prize–winning author of *The Undying*

PENGUIN BOOKS

WARMTH

Daniel Sherrell is an organizer born in 1990. He helped lead the campaign to pass landmark climate justice legislation in New York and is the recipient of a Fulbright grant in creative nonfiction. *Warmth* is his first book.

WARMTH

Coming of Age at
the End of Our World

Daniel Sherrell

PENGUIN BOOKS

for my family, now and then

PENGUIN BOOKS
An imprint of Penguin Random House LLC
penguinrandomhouse.com

An earlier version of the chapter "Correspondence" was published as
"Hunters in the Snow" in *Passages North, No. 41*, in 2019.

An earlier version of the chapter "Fourth Movement"
was published as "A World of Equal Weight" in *Grist* in April 2021.

On page 189, from *Iep Jaltok: Poems from a Marshallese Daughter*
by Kathy Jetnil-Kijiner. © 2017 The Arizona Board of Regents.
Reprinted by permission of the University of Arizona Press.

LIBRARY OF CONGRESS CATALOGING-IN-PUBLICATION DATA
Names: Sherrell, Daniel, author.
Title: Warmth: coming of age at the end of our world / Daniel Sherrell.
Description: first edition. | New York: Penguin Books, 2021.
Identifiers: LCCN 2020051289 (print) | LCCN 2020051290 (ebook) |
ISBN 9780143136538 (paperback) | ISBN 9780525508052 (ebook)
Subjects: LCSH: Sherrell, Daniel—Diaries. | Environmentalists—
United States—Biography. | Environmentalists—Political activity—United States. |
Climatic changes—Moral and ethical aspects. | Environmental justice.
Classification: LCC GE56.S45 A3 2021 (print) | LCC GE56.S45 (ebook) |
DDC 333.72092 [B]—dc23
LC record available at https://lccn.loc.gov/2020051289
LC ebook record available at https://lccn.loc.gov/2020051290

Printed in the United States of America

Book design by Daniel Lagin

Contents

PART I

Correspondence 3

First Movement 31

Outrage 41

Second Movement 89

Loss 101

PART II

Retreat 135

Third Movement 169

Object 181

Fourth Movement 215

Heat 235

Acknowledgments 257

PART I

Correspondence

On April 14, 2018, a civil rights lawyer named David Buckel burned himself alive in Prospect Park. He did it alone, just before sunrise, a brief illumination on a peripheral lawn. A cyclist found his body in a circle of char, though she had to pass by several times to be sure of what she'd seen. Later, she told reporters: it was hard to make myself believe it.

The suicide was well planned, even courteous. Buckel had cleared a ring of dirt around himself to keep the flames from spreading. "I apologize to you for the mess" read a note found by the police in a shopping cart next to the scene. A longer letter had already been emailed out to the press. This was an "early death by fossil fuel," it read. "It reflects what we are doing to ourselves."

I spent most of that day across town in Central Park. I remember it was gorgeous outside and the lakes were all crowded with rowboats, little schools of them flitting back and forth behind the curtain of the willows. I found a perch on top of a small hill and watched the loop road swell with people. Somewhere out of sight, a stoplight was releasing them in pulses: the tourists in their carriages, the cyclists, the loping Rollerbladers. They passed quickly and suddenly, then a beat of empty road, and then the unseen light changed once more, presumably, and the next wave came streaming past. The scene reminded me of a Bruegel painting I'd first come across in a history textbook: a landscape of a village in winter, painted from atop a nearby hill. In it, you can see hunters and woodcutters going about their business, ice-skaters crisscrossing a pond, chimneys smoking in snow. According to the textbook, this painting was meant somehow to delineate the beginning of the Renaissance. As if all it took was a small vantage, the right flow of people, to funnel the whole historical watershed.

After a while, I fell asleep in the grass, and when I woke up the temperature had dropped and the picnics dissipated. The few people still out seemed in a hurry to get home. I walked back through the park toward the East Side, past the closing museums, past the expensive boutiques that mimicked the museums, single handbags underlit in glass display cases. Then down the stairs to the train, which I took back to the Bronx. It was only once I stepped into my darkened apartment that I saw the news from Prospect Park, glancing past it on my phone and then scrolling slowly back up, registering what I'd read.

What struck me even more than the tragedy—and it did strike me, a slow onset, so that I failed to make dinner that night, and eventually, at a loss for what to do once I finally tore myself from the screen, went to bed without ever having turned on the lights—was how quickly the event, this flicker of violence, was subsumed once more into the general mill of the park. Was forgotten, essentially. Beyond the cordon of police tape, the newspapers reported, the barbecues continued as normal, the corporate kickball games resumed. Participants in a charity walk strode industriously by in matching purple T-shirts, which predicted, in cursive quotes, that an end to pancreatic cancer was at hand. Wage Hope, the shirts read. The moment had rolled on, in other words. And I'm only being partially rhetorical when I ask you: What else could it have possibly done?

Afterward, I felt irrationally like I should have been able to detect some ripple when it happened, a subtle shock wave passing from his park to mine, like a bell tolled to part one hour from the next. Undoubtedly the news alerts had been piling up in my pocket, but I'd set my phone on silent and so hadn't felt even those regular vibrations I'd grown accustomed to associating with tragedy. While the man burned—the flames carbonizing his skin, then evaporating his blood—I hadn't felt a thing. It had been a beautiful day, and as I said, I'd spent much of it asleep.

Several days after the immolation, I took an afternoon walk with my mother. We were strolling in circles around a smaller park near my apartment, a third park, St. Mary's, this one less manicured than the other two. The cracked paths bristled with weeds that had sprouted eagerly at that first whisper of spring and then died once again in the ensuing cold snap. They looked brittle now, almost burnt. Apropos of very little, I had told her about the suicide and how sad I was about it. Which was not, in that moment, entirely true. In fact, I was often pathologically adaptive to news about the Problem, and that morning had woken up feeling completely fine, no longer able to access the pain I'd felt just days ago, like I'd stepped out of a room and had it lock behind me. My hope was that invoking the word "sadness" would somehow resurrect the emotion for us both. That I could cast the word like a spell and have it conjure between us an instant commiseration, obviating language altogether, which often struck me as plainly inadequate to any real consideration of the Problem. Though when this didn't work—and it never did—I'd fall back on the usual bromides, letting them drop lamely into conversation. "I feel overwhelmed" is what I said on this particular occasion. "It's such a tragedy, and everyone's already forgotten." It was true, the news cycle had moved inexorably onward, though I still had a few emails at the bottom of my inbox with subject lines like "Rest in Peace," or, in some cases, "Fwd: Re: Rest in Peace."

My mother buried her chin into her scarf, listening. "He must have been kind of crazy," she said, sounding apprehensive, like she wanted to hear me agree. We walked in silence for a moment, past an empty playground, a row of old oaks. In the distance some men were playing soccer, and we watched the ball arc high above their heads.

I didn't think he was crazy, I told her finally, looking down at my feet. From what I knew his life had been normal, even honorable. He'd spent decades winning legal battles for LGBTQ rights before retiring and turning his attention to the Problem. In his last years he'd founded a large-scale composting program, sequestering more and more carbon even as he watched global emissions tick upward. Reading about his death, a part of me had understood exactly where he was coming from. I'd often thought about it myself, I admitted—staging the perfect self-sacrifice, something to drive a lightning rod through the public discourse, a suicide that would finally make the Problem personal. Of course it sounded embarrassing when I said it, messianic and delusional. And I chose not to even mention the specific and melodramatic scenarios that I had at one point or another entertained in the back of my head. Climbing to the top of a smokestack and plunging in, for instance. Or monitoring the weather for the next freak hurricane and then waiting on the beach as the rain picked up and the last bungalow renters pulled out of their driveways until finally the flood would come and wash me far inland and then back out to sea. I'd even run through some of the logistics, like how I'd need to bring bolt cutters in case the smokestack had some sort of grate on top. Or how it would help to rig a GoPro setup before the storm hit, in case the TV cameras weren't there to capture the moment I went under.

"I can't believe you're telling me this," my mother said, lifting her head from her scarf so that we both had to stand there and look at each other. The sun was beginning to set, and we could hear shouts from the field behind us. Someone had scored a goal.

"I'm not depressed," I told her, which was true. For a twenty-something who spent an above-average portion of his time thinking or trying not to think about the end of the world, I was, to my surprise and almost frustration, basically happy. "It's just occurred to me as a strategy," I added, as if this clarification might offer some comfort.

"Don't you ever, ever do anything like that," she said. "It wouldn't do a thing except destroy the people you love."

In the unseasonable cold, burrowed down into her coat, she looked genuinely worried, and I suddenly regretted saying anything. Because of course I knew I agreed with her. It wouldn't do a thing; it was bad strategy. The fire in the park had proven exactly this point: even if you bowed definitively and dramatically out, the Problem would simply go on without you. I promised her I wouldn't, and it was an easy promise, but afterward I felt a new and this time unprompted sadness. Without the possibility of appeal to that ultimate recourse, there was little left to buffer this feeling that I'd been trying to outpace for months: that maybe it didn't matter what you did, maybe there were just no strategies left to us at all.

It was around this time that I began writing to you about the Problem. At first it was only semiconscious, a straw I would grasp at to fend off hopelessness. I'd tap my anxieties into a note in my phone, then click it shut once the thought had petered out. It always felt absurd, trying to confide in you, something I would've been embarrassed to admit to anyone. But it also brought a strange kind of solace: As if I were fortifying the possibility that you might one day exist. As if, in addressing you, I could make you more and more real.

This was after college, in the years I first lived in New York. During that period, I would sometimes accompany my grandmother to her synagogue in Far Rockaway. It was an Orthodox shul, though members referred to it simply as "black-hat," a term I preferred for its note of witchy intrigue.

We would sit apart from each other, my grandmother in the women's section, which was separated from the men's by a one-way screened partition. Through it she could watch me pretend to pray, flipping through the pages in my siddur, my train of thought curling into a little ouroboros of distraction. Around me the other men wore dark suits and tzitzis, occasionally a streimel lined with fur. They prayed in silence, shuckling and mouthing the words until the cantor broke abruptly into song. The singing was always disorganized, entirely absent the harmony of a choir, and those men who finished first would meander through the pews, pulling each other aside for hushed and familiar conversations.

I knew that—invisible behind the screen—my grandmother was praying for you, praying that I'd get married and bring you into the world, and that after services let out she would take me by the arm and point out any number of devout and eligible young women with whom this goal might be accomplished. So I paid particular attention to the rabbi's sermons, hoping to arm myself with another topic of conversation that might steer us away from you, from all the uncomfortable questions you raised.

The rabbi's sermons often concerned the mashiach—when he would come and what we could do to hasten his arrival. The mashiach was meant to usher in olam ha-ba, the coming world, though there was much debate about what this world would consist of. "There is there neither eating, nor drinking, nor any begetting of children, no bargaining or jealousy or hatred or strife," said the Babylonian Talmudist Abba Arika. "All that the righteous do is to sit with their crowns on their heads and enjoy the effulgence of the Divine Presence." The second-century sage Rabbi Yose HaGelili pictured something darker and stranger. "The souls of the wicked," he said, "will be slung away in the hollow of the sling." The Sanhedrin claimed simply that all those who have died will be resurrected. And the book of Isaiah skirted the question entirely. As for the world to come, so it is written, "eye hath not seen."

I agree there may be no use here in my speculating on the nature of the world to come. I mention all of this just to say that on some level I was already familiar with the eschatology of the Problem. I knew what it meant for fate to be dangled just beyond reach, like a carrot on a string. How it might feel to wait a lifetime for something immense and uncertain to finally run its course.

TO GET TO MY GRANDMOTHER'S I took the subway out to the end of the line, where the tracks run parallel with the ocean. The trains clattered past the beachfront apartment blocks, but inside the passengers observed a strict and definitionally unspoken pact of silence. Almost no one talked, and few people made eye contact. They looked down at their phones or out at the water. Sometimes panhandlers and proselytizers broke the pact, but even during rush hour it was like they were speaking to an empty car.

By the time I moved to New York, it had already become clear that the subway system was falling apart. Components were coming loose and connections weren't being made. Old engineers and signalmen were being brought out of retirement, tapped for their arcane knowledge of a certain rusted switchboard at a major junction in Queens. The governor wasn't investing the money needed for an overhaul and spent much of his time trying to blame the mayor, who had no jurisdiction over the matter. But though the infrastructure was outdated and the politics dysfunctional, it was a storm that had really pushed things over the edge. Incongruously, the storm was named Sandy. Some official body gave a name to each successive storm in alphabetical order, restarting each year at the letter *A*. Whoever it was tended to choose sunny names from the Greatest Generation: Sandy, Harvey, Irma—names you might expect to find in a South Florida retirement community, the sort of community that, ironically, was also highly at risk from the uptick in storm severity precipitated by the Problem.

Sandy had struck Far Rockaway in late 2012, once most of the alphabet had already been taken. At the time, I was in my final year of college, hours away from New York. I spent the night texting family members and frantically refreshing my browser, trying to construct from out of the random scraps of news some semblance of the actual scene. How the wind sped up and the trees bent double. How the awnings on my grandmother's apartment building snapped against their tethers. And how everyone inside watched the storm approach on TV, its vortex rendered in vivid green, a juggernaut lumbering toward the coast. They watched until the sky was dark and the rain sounded urgent against the windows, like it was knocking to be let in. They watched until the wind cut the power and extinguished the TVs and they could no longer track where the storm was or what it was doing. Only then, in the blackout of its own making, did the storm make landfall.

The Atlantic crashed through the community, scouring kitchens and basements, blocking off roads. The water ripped up stop signs and buoyed parked cars, stopping just short of the synagogue where for years my grandmother had prayed for the coming of the mashiach. Across the neighborhood, thousands of people were forced to evacuate their homes, leaving hastily as the storm worsened. My grandmother returned the morning after to a darkened apartment, the food spoiling in her fridge. I tried to call her from my dorm room, but the storm had severed her phone lines.

I remember feeling a sense of muted shock: Something long-anticipated was finally happening. Not just happening, but happening to us. The Problem had jumped out of the screen and into my grandmother's living room. It was a moment I knew I'd been bracing for. Even still, it was hard to make myself believe it.

The day after the storm, the subways were a mess. Many of the tunnels were still flooded with ocean, its salt eating into their walls. The images online looked like how I pictured the River Styx, a watery cave disappearing into complete darkness, just big enough for a Central Park rowboat.

In the years following, I watched the delays get worse, the

rush-hour crowds more packed and frantic. This deterioration coincided with a proliferation of signs and apps that told you the exact amount of time until the next train, so that as the system's reliability decreased, its predictability increased. It felt like this indicated a subtle shift in our relationship to time, like we were no longer content to trust its progression but, like jealous spouses, felt better when we knew what it was up to. Though of course, knowing the wait time had no impact on its duration and, I suspected, only increased our impatience.

As the trains continued to fail, there was also a change in the silence among the passengers. Less rote somehow, more apprehensive, like a conversation was forever on the verge of breaking out. Coming home late at night I would sometimes see men my age talking to their reflections in the windows, mouthing the lyrics to the private songs playing in their ears. You could watch their mirrored selves, jerking and ducking against the darkness of the compromised tunnels. Then the train would pull into the station, and their own image would disappear from the window, replaced by a well-lit stranger waiting impatiently for the doors to open. There was a sense that everything was just barely hanging together.

This fragility made the arrival of every train feel like a miracle. Even after twenty minutes, that one had come at all was a testament to the persistence of some much larger house of cards, balanced precariously just out of sight.

———

DRIVING EVERYTHING—THE STORMS and the trains alike—is the instrument of David Buckel's suicide. Fossil fuel is everywhere, though rarely seen. It hides itself in pipelines, snaking under our streets and through our walls, arcing across remote hills and tracing the bathymetry of the seabed. Even at the gas pump, the point where it seeps furthest into our conscious attention, oil passes unseen from nozzle to tank, leaving behind only a very polarizing smell.

The nature of fossil fuel is essentially ambivalent: an invisible pervasion that powers all things and will also, inevitably, destroy them. It is produced by drilling, extracting, and refining those delicate ferns and planktons that millions of years ago were ground to a sludge by the planet's crust. The sludge gets processed into coal, gas, or petroleum, then shipped off to power plants where it is converted into energy. In this way we power the present almost exclusively by burning the remains of the past. Unearthed, our history surrounds us, dissolving through the air, until its ubiquity comes to look very much like the future itself.

Another way to put this is that we are involved in a kind of trans-geologic grave robbing, in light of which, the Problem can rightly be seen as a haunting.

During the time of the suicide, I was working for NY Renews, a statewide coalition that was trying to contain the haunting. The coalition had been built to reflect the diversity of people it was attempting to protect. There were pastors and students, unionized nurses and career energy wonks. There were upstate farmers who'd fought to defend their land from fracking. There were communities of color—from Brooklyn to Buffalo—who'd seen power plants erected in their neighborhoods, who'd been forced for decades to breathe disproportionately polluted air.

All told, over 150 labor unions, community groups, and environmental organizations had banded together in the effort, and it was my job to make sure everyone played nice and stayed on task. Our collective goal was to pass a package of two bills: The first would require the state of New York to eliminate greenhouse gas emissions by 2050, including from electricity production, transportation, and buildings; prioritize funding for low-income communities; and put in place labor standards to ensure the creation of good jobs in the renewable energy economy. The second bill was simpler: it would help fund this massive leap forward by taxing the companies most responsible for the Problem.

The trick with the work was not to think about the sheer length and delicacy of the causal chain we were trying to set in motion. Every meeting, rally, and press conference we organized was aimed at bolstering support for our demands. If we could build enough momentum, the thinking went, then we could marginally shift the governor's political calculus on the issue, a calculus complexly derived from his polling data, the priorities of his top donors, and the proximity of an election. If we sustained our pressure and the other variables fell our way, then the governor might be compelled to pass unprecedentedly ambitious and far-reaching legislation to address the Problem. And if the governor passed unprecedentedly ambitious legislation in New York then, given the size and influence of our economy, other states might follow suit, at least in the immediate region. And if other states followed suit, then a new standard could be set to help prescribe the actions of the next federal administration. And if the next federal administration took up the torch we'd lit in the Northeast, then this would hopefully coincide with a similar strengthening of resolve from every other country in the world. And if such an upward spiral could be set in motion, then at least on paper it was theoretically possible for us to contain the Problem enough to prevent the level of fires, droughts, and floods that would render sustained civilization impossible.

The point being that this is sometimes what hope looked like that year, and so I couldn't always look that closely.

What I looked at instead was my phone. The job, like many jobs, required it. My phone was where I took calls, wrote emails, checked the news out of Albany. When it wasn't in my hand, it sat expectantly in a pocket, pressing against my thigh through a thin layer of cotton. It had the weight and polish of a river stone, with fine cracks splintering the screen. On its back I'd fitted a case with overlapping green, white, and blue chevrons, like a stylized forest of evergreens.

My phone, as I mentioned, is where I began writing notes to you. Shuttling between meetings on the subway, I would suddenly think of something I wanted to tell you and reach reflexively into a pocket. My phone's resting face showed an anachronistically vibrant coral reef, a picture I'd chosen years ago and never replaced. When I pressed its circular button twice, the phone would show me everything I'd been doing the last time it had been in my hand: checking my bank account, for instance, or remarking on the birthday of an acquaintance. Each of these activities had its own pane, and these panes stood one behind the next, receding into the screen like hills into the distance, each one a shade darker. To write you a message, I'd place my thumb on the foremost pane and fling it to the right. The rest of the panes would career after it and then slowly decelerate until, if my thumb had applied the right amount of initial force, they would come to a halt on the pane where I composed notes, which had been designed to look like a page of blank paper.

Afterward, I'd scroll over to my work emails, which the thought of you always prompted me to check. Inside the subway tunnels I didn't get any service, so I would wait for the train to pull into a station and then, as the doors opened, quickly swipe my thumb downward to replenish my list of emails. At the top of its face, my phone would display a little gray daisy, rotating its petals to quell my impatience. And just in time, as the doors were closing, a new block of messages would appear, each one marked with a dot as blue as a berry.

I am trying here to evoke my phone like a landscape, a territory through which I regularly passed. This is an exaggeration, though maybe not much of one. As was true for many people, my phone was a nearly ubiquitous backdrop, and in front of its screen I performed many of the activities that made up my daily life. I used it to exchange money, to find dates, to find food, to listen to music, to talk to my doctor, to talk to my therapist, to check the weather, to check my calendar, to take photographs, to take videos, to masturbate, to meditate, to find my way around, to find out what my friends were doing, to find out what their friends were doing, to see in the dark, to track trains, to get a lift, to look up words, to search for any information at all really, and occasionally to call my grandmother on her landline. Sometimes, I bid it speak and it would read out recipes, or coach me through a workout, or wake me up in the morning by imitating the sound of a wind chime.

I made various attempts at containing its influence: silencing it so I wouldn't hear its buzz, banishing it to my outer coat pockets so I wouldn't feel its weight, changing its color scheme to the matte grayscale of newsprint. I even left it at home sometimes, on the rare weekend when I wasn't expecting calls from any of the 150 organizations in NY Renews.

None of this proved particularly successful. And the more strands of my life that were routed through it, the more I came to associate my phone with the nature of the Problem itself. Because it wasn't this simple, but this is how it felt: That as the world around me began to fall apart, a new one was blooming outward from the palm of my hand. And as the former reeled from floods and wildfires, a retreat to the latter seemed not only more tempting but increasingly more viable. Even the storms themselves were routed through my phone now: At the first hint of flooding, every screen in the city would brighten simultaneously with alarm.

Over time, as the news got worse, the notes on my phone began cohering into this letter. This hadn't been my intention. I'd stumbled into writing them, a private and haphazard stab at processing. But the more I addressed you on the page, the more I came to feel responsible for your existence beyond it. I realized that if I was ever going to actually start a family—if I was going to move you from the flourishing world in my palm into the collapsing world at hand—then I'd owe you an honest account of why. Not just the decision, but its context, the whole story: what I thought about and what I read, how I felt and how I was numb, where I found faith and where I harbored doubt. And how difficult it was to hold out hope, to keep you viable against the rising mercury. Not because I didn't love you but precisely because I did, because I do. This is the point of my letter—maybe of any letter from parent to child. To show you that love. To show you where it is you've come from.

———

HERE, THEN, is an important piece of context, something I'd like you to know up front: Occasionally, amid all the storms, I'd feel grief. The grief had its own weather. It would come down like a squall, momentary and encompassing, impervious to forecast. This was a weight I kept private mostly, unsure of whether or how to share it. Even with my closest friends, discussions of the Problem tended to stumble into the arid gully of knowing commiseration. "We're so fucked" is what we let ourselves say, on the rare occasion the conversation wasn't quickly diverted into lighter terrain.

But alone, it rose up in me like a whale from a depth, almost invisible until the moment it breached, water streaming from its flanks, the most powerful thing in the world. And like with a whale, the breaches seemed to come at random, when I least expected them. I cried about it on line at the grocery store and in the bathroom at parties and by myself in the shower. Never loudly, just a few tears, messy and quickly stifled.

Sometimes I'd share my grief with my partner at the time, though this wasn't always intentional. I remember one night she came home in a gloomy mood and started telling me about something stressful that had happened at work. I really did think I was listening, until halfway through her story I began tearing up out of nowhere. "You can't make everything about the Problem," she said, frustrated, aware of what was happening. "I'm allowed to be upset about other things." She was right, of course. My relationship to the Problem often does this, turns me into an asshole, myopic in my grief. And still I wanted to say: Everything *is* about the Problem! The Problem is about everything!

Other times, I could read about global crop failure and drowning cities and not feel a thing. I remember once, on a subway ride home from work, pulling up an article by Elizabeth Kolbert. It was a piece I'd been meaning to read, a feature on the prospect of carbon removal technology. The thesis was basically that there were no longer any paths to avoiding catastrophe without the large-scale deployment of atmospheric carbon removal, and that large-scale atmospheric carbon removal was for all intents and purposes impossible. The article had been published the previous year. Putting away my phone, all I could muster was a dull outrage that it'd taken me months to find out that the end of the world had been convincingly proclaimed on page seventy-two of the November 2017 *New Yorker*. I walked back to my apartment as usual, stopping by the crates of produce put out for sale along the sidewalk. I bought a cactus paddle, and when I got home, I cut its spines out and cooked it with eggs. Nothing else happened that evening that I can remember. It was perverse. Even now, as I write this, I don't feel any sadness, I am merely telling you that I have.

———

GROWING UP, this numbness was the rule, so ubiquitous as to be undetectable. Our family lived in a leafy suburban town tucked into the elbow of a notoriously polluted river. The river was wide and brown and, it was understood, unfit for swimming. We never did, not even on a dare, though we drove over it frequently to get to the AMC megaplex and, beyond that, to my parents' research campus.

My parents were—still are—professors in the hard sciences, their offices located in adjacent buildings at Rutgers University.

My mother studies how the human body metabolizes fat. My father is an oceanographer whose research concerns the chemical interactions between the oceans, the atmosphere, and marine life. He was in Greenland, in fact, when my mother was pregnant with me, spent the second trimester drilling cylindrical samples out of the ice cap. These samples could be read for any number of things: the composition of past atmospheres, the history of regional air temperatures, the dynamics of accumulation and ablation on the ice cap itself.

Picture him there, in his fluorescent parka, a mote of orange on a giant lens of white. The wind is sharp, and the sun is so bright he wears dark glasses to shield his retinas. At night, the temperature drops to twenty degrees below zero. The glacier outside his tent is moonlit and inert, flattened beneath a colossal silence. Over the next thirty years, the Greenland ice cap will start to melt: gradually at first, and then with alarming speed. Threads of water will snake across its face, pooling into crystal wells, carving small rivers with banks of ice. But for now, this process has only just begun, and the ground beneath his boots feels cold and solid.

Among my father's colleagues, there is an understanding that a big thaw is possible, maybe even imminent. But, worrying as they are, these conversations remain theoretical, percolating quietly at the margins of the literature. Do they spark anything in my father, make him consider the child he's about to have? Could this question have occurred to him, to anyone, in 1990?

Regardless, he returns home to New Jersey, and three months later I am born into a world that's begun imperceptibly to melt. I am a happy child, and we are, for the most part, a happy family. My sister arrives two years after me, and soon we're thick as thieves. There are pictures of us standing on our front steps in a blizzard, bundled beyond recognition; or seated absorbedly in our high chairs, eating dry Cheerios one by one. As we grow up, my father's research switches poles, and he begins taking long trips to the Antarctic. He is away for two, sometimes three months at a time, skirting the continent in an icebreaker, taking carefully plotted samples of the Southern Ocean. I am proud of him, and also decidedly jealous. He tells us about taking a Zodiac out to a passing ice floe, the scientists and crew all avoiding the edges, where fourteen-foot leopard seals have been known to leap out of the water and grab a leg. He tells us about days spent at a British research station on the Antarctic Peninsula, cold games of soccer played on a gravel runway.

Beneath these stories, the Problem lurks like a subtext. Few places on the planet are warming faster than the peninsula. The sea ice that lines the coast—like one vast, floating carpet—is beginning to contract. The ice sheets that dominate the land are sloughing off huge quantities of liquid, cleaving increasingly large chunks of themselves into the sea. This influx of fresh water has upended coastal ecosystems, precipitating changes in plankton populations and nutrient distributions that shudder their way up the food chain.

But though these facts might have primed our family to talk about the Problem, we somehow never did. To the extent it existed for us at all, it was as a passing referent, a troubling but arcane detail of my father's professional world. And though he spoke often about his work, he rarely inspired any gloom. He'd grown up building bicycles and toying with car engines, and had come into oceanography at least in part through his enthusiasm for tinkering. His job allowed him to design complex and delicate new sampling instruments, ten-thousand-dollar machines that were lowered by winch into the frozen ocean. The existence of the Problem, then, seemed almost incidental, a heavy fact that aligned poorly with my father's curiosity and excitement, his natural relish for the world.

For her part, my mother half dreaded his trips to Antarctica. Not for what they'd reveal about the Problem, but because they meant attending to two teenagers alone, not to mention keeping up with her own research. By then my sister and I were in high school, with the inklings of our own lives. My sister—fiendishly smart and prone to anxiety—had begun medicating herself with weed, sneaking down to the empty lots on the edges of the toxic river. I spent most of my time overstudying for tests and obsessively cutting weight for the wrestling team—two arenas in which my capacity for effort correlated neatly with my desire for control.

When my father returned, we could tell that he'd missed us, felt bad for being away. As a kind of consolation, he'd show us long slideshows from the trip, all of us squeezed together on the couch in front of his laptop. There were videos of emperor penguins, hunched and dour as sentries, and of little Adélies skating on their bellies across the ice. There were pictures of icebergs that looked like statuary, abstract and newly calved, their massive bodies sitting low in the water.

Sometimes during these slideshows, I would feel my dad wanting to talk about the Problem. "It's not the same continent I visited a decade ago," he'd say absently. But none of us knew how to pick up this thread, where to take it. We just sat there huddled on the couch, watching the penguins flit across the screen. They slid face-first across the melting floes, then dove straight into the water.

My father, for his part, remembers one night when we did broach the topic. I was around eight years old and must have seen someone reference the Problem on TV. I asked him what it was, and he remembers worrying about how much he should tell me. He answered honestly, trying his best to sound reassuring. But when he finished, I grew indignant and started to cry. How had he not told me about this before? he recalls my asking. We could've been doing something about it this whole time!

Unlike my father, I have no memory of this conversation. Perhaps his worries were well founded. Maybe it was just too much, too early. Maybe I blocked it out. Or perhaps he imagined it in retrospect, the way our memories can invent scenes that fill the gaps in a story—scenes that, even if they never happened, surely should have.

Either way, I'm not sure it matters. The point is that even with a father who loved me very much, even with a father who was emotionally available, and passionate about his work, and far more knowledgeable on the subject than almost any other parent, it was still hard to address the Problem head-on.

According to my memory, it took almost thirteen more years for the weight of the Problem to provoke in me a commensurate sadness, and even then, it lasted for only about twenty minutes. It was fall 2011, a month before my twenty-first birthday. My father was in Antarctica, and I'd come home from college to spend a long weekend with my mother. I remember the clocks had just been set back, and the early nightfalls still felt abrupt and unfamiliar. We decided to watch a movie and, knowing nothing about it, I suggested Lars Von Trier's *Melancholia*, which I'd been assigned to watch anyway for a class. My mother propped the laptop on her knees and we sat up against her headboard staring at its screen.

In the movie, Kirsten Dunst plays a woman suffering from

melancholic depression, and Charlotte Gainsbourg plays her sister. The plot revolves around the sudden appearance of a new planet in the sky, which scientists have dubbed "Melancholia." Gainsbourg is confident, in line with the prevailing wisdom, that Melancholia will pass close to Earth and then shoot past it. Dunst is convinced it will make direct impact, killing everyone.

On the day of the projected fly-by, the family gathers on the terrace to watch the spectacle. The planet floats like a sightless face, a huge pale mask, over the grounds of their palatial home. Gainsbourg's young son twists a green twig into a circle, which they use to track the planet's progress across the sky. As Melancholia flies past Earth, Gainsbourg watches it grow smaller and smaller in the rudimentary frame, receding into space. She is giddy with relief, though Dunst is listless as a prop, hardly responding to the news. The next morning, Gainsbourg places the twig to her eye once more and is shocked to see that the planet appears to have grown larger. In a panic, she places the circle to her eye again and again, but the result doesn't change: in the pull of Earth's gravitational field, Melancholia has circled back and is headed straight for them.

In the last scene, Gainsbourg, her son, and Dunst share a final meal together. Above them, Melancholia has replaced the sky itself, its cratered surface obscuring all the blue. For the first time in the movie, Dunst is calm—there is a new life to her gestures, as if Melancholia has brought something inside of her into alignment. Dunst tells her nephew not to worry, that they can escape the collision by hiding in a "magic cave." Grabbing her sister too, she leads them to the top of a small hill and constructs a fragile tepee from broken branches, insisting they all sit inside. As the planet gets nearer, each of them adopts a pose: Gainsbourg is bent double and sobbing; her son has closed his eyes, trusting in the cave; and Dunst sits cross-legged, watching them both, a new sense of peace straightening her back and relaxing her shoulders. Then Melancholia makes impact, destroying the entire world.

When the movie ended, we watched the credits scroll across the screen, neither of us making a move to break the spell. I could feel the heat from the bottom of the laptop, the beleaguered whir of its little fan. Then all of a sudden I was sobbing, really sobbing, like the sobs were gripping my arms and shaking me hard enough to dislodge tears. It was so sudden that at first I didn't know what was happening, and so couldn't explain to my mother that the movie had finally induced the kind of feeling I'd always assumed was latent in my understanding of the Problem: this fathomless realization that the world I knew could end, might end, was perhaps in the process of ending.

The feeling reminded me of certain dreams I'd had. In these dreams I would plunge into a body of water and realize I was sinking. Casually at first, and then with increasing urgency, I would thrash my limbs around, trying my hardest to swim. But no matter how hard I kicked, the surface would continue to recede, and the water would grow blacker and blacker. Eventually it would occur to me that I'd exhausted all my options, and that I was soon going to drown. I would feel my mind wrapping its arms around this fact, trying to press it into an alternative formation, squeeze from it some last trickle of hope. And when this failed, the gates of my mouth would open and resignation would come flooding in with the water—a sudden slackening of the muscles, a last-minute shuffling of priorities, the bounds of my foresight snapping back to within arm's reach. Then I would wake up suddenly and, without moving, stare into the darkness above my face, trying and failing to retain this feeling that was so elusive in waking life: the feeling that I could die, was in fact going to die.

But like with those dreams, the feeling induced by *Melancholia* quickly dissolved. I stopped crying and got up without explanation, assuring my mother that everything was okay.

I went to bed that night thinking about Kirsten Dunst's character. It didn't seem like a coincidence that the depressed sister was the one best prepared for the apocalypse, the one who could see it most clearly. It was like those studies showing that depressed people are the only ones who can accurately evaluate their driving ability, the rest of the respondents just assuming they're better than average. In her sadness lay a certain lucidity.

To me this was all a cruel irony: how you had to *already feel* like the world was ending to be able to assimilate the truth when it actually did. And no matter how hard I tried, I couldn't feel it, not consistently. The selective blindness, the congenital optimism— it all felt too hardwired. Which is why, over time, I came to weirdly treasure those fleeting moments, like the one that evening, when the depth of my grief seemed to match the gravity of the Problem, when the gap between fact and feeling appeared momentarily to close. Because even through my devastation, I felt like I was brushing against some truer reality, the empirical one, the world as it was outside and prior to the filters with which I otherwise convinced myself, silently and as a matter of course, that everything was fine.

———

IT TOOK A FEW DAYS after Sandy for my grandmother's phone line to get reconnected. When I finally got through to her, she did not want to talk about the storm. "Unusually bad" was all she had to say on the subject. Instead she wanted updates on the usual. She asked me about school, about when I was coming to visit. She asked me if I had a girlfriend, and I told her that I did. "So?" she said. "When are you going to start a family?" This was the same question I'd been fielding from her for years, that I was entirely used to dodging. But that day the question felt different, heavier, as if the storm had exposed its relevance. In the pause on the line, it felt like I was considering it—considering you—for the first time. And all of a sudden I knew that I really did want to have you and that this conclusion had taken shape prior to and underneath any conscious deliberation.

Already my ex and I had dabbled in that dangerous game of imagining you, though we'd never decided to have children, let alone with each other. I'd found I could picture you clearly, strapped to me with one of those baby backpacks, the kind that rides in front, so that when you fell asleep your head would rest on my chest and your arms would dangle next to mine. I'd had predictable little reveries of pulling socks onto your tiny feet, and wrapping you in a towel at the beach, and pretending to chat nonchalantly with other parents as I nonetheless watched you out of the corner of my eye hanging upside down from the monkey bars.

But standing with my ear pressed to the receiver, these feelings terrified me. I had had no hand in them. They'd appeared fully formed, as if they'd always been there. I pictured the debris sitting outside my grandmother's door, the soggy trash and hunks of plywood. I pictured the milk rotting in her fridge. I pictured my genes like a virus in my body, co-opting it, infecting me with the idea of you.

"Soon," I assured the woman who aspires to be your ances-
tor. I think, but am not sure, that this was a lie.

In the years since then, the storms have gotten bigger, setting
new records for damage and death toll. On the day I began writing to you about Sandy, Hurricane Michael made landfall in the
Florida panhandle, clocking wind speeds of 155 miles per hour.
For days its ghost stalked north up the seaboard, blackening the
sky out the window by which I sat trying to describe the aftermath of that previous storm. And this was what terrified me:
not only the storms themselves, but how their violence could
manifest as just patter on the pane, ideal for writing letters. Already there were reports of whole families swept away, but I
was safe there at the edge of the system: the weakening cyclone
had brought us only a heavy rain, which I could watch as it trel-
lised the glass.

———

AT THE BEGINNING OF *MELANCHOLIA* there's a ten-minute overture
sequence set to an opera by Wagner. Von Trier strings together
a series of static shots filmed in slow motion, the figures barely
moving. On rewatching the film, I've come to notice something
I hadn't seen the first time: amid this opening montage there is
a lengthy shot of that same Bruegel painting, which I've since
learned is called *Hunters in the Snow*. The frame is full of it: the
icy hill and its copse of birch, the dogs and men poised on top,
drawing your gaze along with theirs toward the tiny bustle of the
town and the mountains beyond it. For several seconds nothing
happens. You watch the painting. Then shards of black begin to
fall, obscuring parts of the image, and you realize that the painting is burning, bits of ash flaking off from the top.

The thing about watching a painting burn is that it elicits no reaction from within the painting itself. This shouldn't be surprising but is, somehow. You half expect its figures to revolt against their demise: for the birch trees to bend, the dogs to howl, the hunters to flee or beg. But everyone and everything holds its pose, even as the fire peels back the margins.

———

TEN MONTHS AFTER SANDY, on the first day of the Jewish New Year, I accompanied my mother and grandmother to Rockaway Beach to perform tashlich. It was early fall, and a cold front had killed the wind. The surface of the ocean barely moved. Behind the dunes, a part of the boardwalk was still uprooted and weirdly bent, its wooden slats looking for all the world like the spine of a whale. There were other remnants of the storm: houses we walked by with boards on their windows, faint tide lines still staining the vinyl. But the water itself betrayed nothing, not a hint of what it had been or could be.

Tashlich is a ritual where you're meant to cast out the sins of the previous year so you can start the new year afresh. You do this symbolically by consecrating pieces of bread and tossing them into a body of water. My grandmother had brought a whole-wheat loaf from her freezer, still half frozen, and we stood there silently on the beach, three consecutive generations, tearing it into little pieces of shame and regret.

I wasn't sure what to call my sin exactly, but I dedicated its atonement to you and threw my piece at the horizon. The bread was stale and didn't fly well. It landed in the shallows a few feet away, where the seagulls scooped it up and lurched off, unaware of the karmic implications.

When I finished, I took my shoes off and walked to the edge

of the water. I could see my mother and grandmother down the shore, still clutching their pieces of bread, heads bent in a pose of mourning or of prayer. I stood very still then, waiting for the water to reach me. But the tide, it seemed, was on its way out.

First Movement

It's September 21, 2014, and half a million people have gathered on Central Park West. I've never seen this many people in one place, enough to clog all the side streets, to completely obscure the pavement. "Throng" seems too small a word for what is happening. At this scale, a crowd begins to read more like a migration, like a sizable sampling of the species.

And though we've spent months calling them here, it's still impossible to believe they've arrived. We frankly don't know where they've all come from, but here they are anyway, hundreds of thousands of strangers gathered together on a street that was empty just this morning, save for a few taxis and the long steel barricades of the police.

Across town at the United Nations, world leaders are gathering to discuss the Problem. Our gathering is a corollary to theirs, meant to goad them into urgency and turn their discussion into action. We've billed it as the largest ever march against the Problem, and within twenty minutes of the start time it is clear that this is true by a long shot, maybe even by an order of magnitude. It's been my job, over the past few months, to find ways to spread the word. I've given recruitment speeches at gurdwaras and community boards and punk venues where the stages stink of beer. I've gotten so facile with the invitation that I can tailor it to anyone, bring it anywhere, shrink it to two sentences at the bar or stretch it to twenty minutes for the congregation of a church.

This is my first job out of college. I've been hired by the Sierra Club with a loose mandate to help drive turnout for the march. There are about a dozen of us, young organizers brought on temporarily by one or another allied organization, entrusted collectively with spreading the word. Most days we fan out across the city, talking ourselves blue in the face but also trying to listen, to have real conversations about shared concerns. For the most part, we fail. We bring crews of volunteers to subway stations and music festivals, spend whole afternoons handing out flyers for the march, but few people want to stop to talk about the Problem. We'll see them see us from afar, angling their steps to ward off our approach. Sometimes they'll grab a leaflet without a word and we'll watch them crumple it into their pocket a few steps away. I try to remind myself that this is understandable. At base we look like all the other touts, the ones from strip clubs and tax prep agencies and the Humane Society. On the wrong day or in the wrong mood, I would walk by me, too. But still I feel embarrassed, trying to peddle my concern for the Problem and having it rejected again and again. I also feel embarrassed that I'm embarrassed, as if the urgent importance of fighting the Problem should by itself be enough to let me hold my head high, to approach each new pedestrian with an unflinching sense of purpose. This is the posture I try to feign for the benefit of my volunteers, anyway, though a part of me cringes every time I have to walk up to a stranger and muster my enthusiasm.

The few people who do stop, though, tend to engage with an almost confessional fervor. It's like they've been waiting for the conversation, for any opportunity to express the anger or panic or incredulity they've been nursing privately about the Problem. Sometimes this is gratifying, though it often feels sad—that even in 2014, there would be so few outlets for talking honestly about the Problem that people would unburden themselves to random twenty-year-olds on a subway platform.

Amid these scattered conversations, we have a hard time predicting how many people will show up. Until today, we had been hoping for a hundred thousand—less than a quarter of the number of people who have actually arrived. Many of them are carrying banners, some long enough to require ten sets of hands, and they slump and flap wherever they've been dropped, twisting like currents through the crowd until again they're pulled taut, all the letters visible. Screen-printed flags on poles of bamboo drape in from the sidewalks and hover above the street, waving their images of children or planets or thermometers. There are also giant white birds, elaborate puppets, and they swoop and dive over everyone's heads, symbols of what it's unclear, though this ambiguity lends them drama, makes them capable of representing practically anything: peace, fear, beauty, rage—everything unspoken we've brought with us to the march. Some of the banners are shaped like parachutes, and people gather around to lift them into the sky, their messages doming above the surface of the crowd. The entire avenue is like a river of ink and people and fabric and noise, like a dam has broken into the canyon between skyscrapers. And it's not hard to picture us this way, each a tiny droplet filtered from who knows where in the broader catchment, seeping through our personal soils of doubt and resolve to arrive here, en masse, a torrent borne of trickles. As it swells our new river cleaves Manhattan, an island already surrounded by rivers, which rise higher every year. We are in this way a preemptive flood, an inoculation against something literal and imminent.

When we finally start moving, I climb up one of the parapets on the stone wall enclosing the park. Its top is faintly pyramidal, so I have to stand bowlegged to look out over the march. I squint hard but I cannot see the end of it, just a carpet of marchers in every direction. They let off a steady roar that is simply the accumulation of their normal voices. As the river streams past I struggle to focus on any one person—it's like trying to trace a bubble or a twig in a stream, individuals resolving back into the general flow.

The organizers have been outfitted with fancy walkie-talkies, but once the march begins it is out of our hands; there are no more directives to be given, the crowd just lurches off by itself, loosening its shape and funneling down the barricades. People amble and dawdle, they double back, they duck away into McDonald's bathrooms. They stand tiptoe on the curb taking videos on their phones. In our earpieces there is just a garble of crosstalk, diced up by static, everyone trying to locate themselves and one another in the ecstatic flood.

I have the keen sense while we march that history is being made, though not exactly in the grand manner of watersheds and turning points, more just in the fact that all of a sudden I can picture the whole scene from your vantage, the same way I've seen choppy films of Armistice Day or suffragette rallies, all those grainy bodies mobilized to the tune of some now-obscure set of circumstances, and I can see how, from out there in the world to come, we might very well appear sort of naive, at least in the fundamental sense that we do not know what is going to happen. That you do and we don't, and you can recall us in our ignorance, looking back through the lens of what's transpired.

In that moment of temporal alienation, the thing that seems strangest to me is the walking itself. I can imagine you thinking this. Why did they march at all? What did walking from Fifty-ninth Street to Thirty-fourth have to do with the Problem?

I can give you the usual explanations about social movements and the demonstration of power, though probably the same sort of power could be projected standing still. Maybe the marching is meant on some mimetic level as a literal embodiment of prog-ress, the feeling of starting somewhere and ending somewhere else, as if to plot our political hopes onto the grid of the city. Maybe it's therapeutic, too, like any long walk devoted to a sin-gle train of thought (processing via processing).

Or I can put it to you this way: marching is simply one means of telling a story. That weekend, there are many stories orbiting the UN conference, all of them vying for preeminence. There is the story told by the denialists, in which the Problem isn't hap-pening. There is the story told by many governments, in which the Problem is happening, though not urgently enough to justify any immediate drawdown on fossil fuels. There is the story told by the fossil fuel corporations themselves, which have wedged their way into the conference in order to tout their commitment to fighting the Problem while watering down any language that might hold them to it.

To lift our narrative above this cacophony, the organizers have built its arc into the structure of our march. At the front

are communities whose marginality means they are already being sacrificed: places where the corporations have ripped fuel
from the soil or poisoned the air by burning it; places in the city
that are still struggling to rebuild after Sandy. Mostly these are
people from poor communities, indigenous communities, Black
and brown communities—people for whom the Problem is already a matter of life or death, who lead the resistance as a matter
of survival. "Frontlines of crisis, forefront of change," their banner reads. After them comes the youth bloc, the one I belong to.
Young people are also being sacrificed to the Problem, though in
a different way, more temporal than spatial. Still our banner car-
ries a note of hope: "We can build the future!" it says. And behind
us come the green businesses ("We have the solutions!"), then
the fossil fuel fighters ("We know who is responsible!"), then the
scientists and faith leaders ("The debate is over!"), and then
finally everybody, an undifferentiated mass of neighbors and
union members and gardeners and grandmas ("To change
everything it takes everyone!").

Here, then, is the story we write by marching, a story that
doesn't look away from injustice, a story that gives agency to the
people and trusts them with it. It is a democratic story, a plural-
istic one. It finds strength in difference without inflaming division or imposing similarity. It finds hope in crisis without caving
to blind optimism or deflating in despair. The story has its flaws,
which our opponents are quick to point out. It milks pathos,
skimps on nuance. But unlike theirs, it has the advantage of being true. We endorse its truth with our feet, walking a combined
800,000 miles, enough to travel thirty-three times around the
small planet we've painted on our signs.

If nothing else, marching is work. You sweat, you shout,

maybe your legs get a little sore. This is one way to take ownership of a narrative, at least in the Lockean sense that putting work into something makes it yours. So we move to seize the plotline, to take it back from those who would soften or obscure it.

And at least on that day, our work works—the story is ours. On the front page of *The New York Times*, right there above the fold, are four pictures of the rally. We wake up to them sitting in coffee shops and bodega windows, the headlines as triumphant as we could have asked, though what this amounts to is not immediately clear. Inside the United Nations, we hear, the sympathetic leaders made reference to us in their speeches. The secretary general himself linked arms and joined the front of the march. But stories, we know, do not move like marches. They do not travel neatly toward a single destination. Over months and years, the story we've tried to tell will percolate outward into the zeitgeist, tipping some dominoes, putting its finger on certain scales. Even today, I cannot tell you exactly what it's done or where it's been. Every march is an act of faith in this way: you have to trust your story will braid into history, even if you'll never be able to tease out its thread.

I still remember one of those pictures from *The New York Times*. It showed a little girl on her father's shoulders, lifted up above the throng, holding a sunflower made of cardboard. There was so much hope in that picture, so much tenderness. And looking at it I felt suddenly like I was looking at a picture of you, not a resonance but an actual likeness, like the little girl really *was* you. The sentimentality of the thought embarrassed me, though it'd been completely involuntary, like I'd witnessed an apparition. The idea of you grafted onto the image of her until it felt like I was looking simultaneously into the near past and the distant future—your future, that is, and by extension my own. And the thought struck me of how nice it would be to march alongside you, to be able to fight and mourn together. I pictured lifting you onto my shoulders, holding your hand, watching you run ahead into the crowd. I pictured you alive, in defiance of the worsening projections, in defiance of every executive and premier who'd just as soon see you dead. I knew that I loved you then, and that love could feel a little like fury.

Midway through the march there is a moment of silence for the lives already lost to the Problem. The organizers have planned it, though I privately doubt it will work. It seems like an impossible thing to coordinate with half a million people; I've never witnessed mourning at that scale.

And yet here it is, a colossal hush sweeping up the crowd, like something is sucking all the air out of the streets. No one knows where it's come from or who's given the signal, but you can actually *hear it* approaching from blocks away, though of course there is nothing to hear. When the silence reaches them, people catch their breath and stand still. The quiet is heavy, almost corporeal. Not the absence of sound but the presence of something that cannot be spoken.

The march ends on a wide street bordering the Hudson, and when we finally arrive your great-grandmother is waiting for me. She's come to show her support, though she is too old to do any marching. I find her seated on the little cushion of her walker, which she's parked next to a fire hydrant on the curb. She gives me a hug, but we don't talk much. We just sit there watching until the ends of the march have trickled away and the street is empty again. Uncharacteristically, she doesn't bring you up once that day. I think we both know you are already there, a ghost of the march, fading toward the river with the rest of the crowd.

Outrage

Though it'd formed a distant backdrop to much of my childhood, it wasn't until college that anyone asked me to do anything about the Problem. This was in 2009, the beginning of my freshman year at Brown. I remember walking across the quad and being approached by a woman named Emily, who wanted to know if I would make a call to the United Nations. She and a few others were trying to generate as many calls as possible, urging international leaders to take ambitious action at the upcoming Copenhagen climate summit. I was only vaguely aware of what the summit was, or why it was happening, but I knew Emily through mutual acquaintances. She was tall and trenchant, and—on the few occasions we'd met previously—had struck me as someone I wanted to befriend.

Behind her, a small foldout table had been set up and arrayed with pamphlets. Several people—almost all of them women—stood around holding their cellphones to their ears, apparently on hold. "You should come help!" Emily said. She had the tongue-in-cheek enthusiasm of someone overcoming her own discomfort, though I sensed too that she really did believe I had a responsibility to help.

At the time, it seemed a little preposterous to me that a group of teenagers in Rhode Island could just call the United Nations and make a demand. Still, I felt implicated by the request. I knew in theory that the Problem was important, but this importance had always seemed to place it firmly out of my reach, in the hazy firmament of ministers and diplomats. I had somehow never really considered the possibility that I could—and therefore, inescapably, should—try to intervene myself.

Taking a pamphlet from the table, I tapped in the number she gave me and waited to hear a ring. I do not remember who I spoke to that day, nor indeed if I spoke to anyone. Most likely I left an awkward message at the dead end of a phone tree. What I do remember is feeling a small thrill: the realization—obvious though it'd always been—that action was possible, that there were other postures available beyond passive concern.

Sometimes people ask me how I "got into activism." I never answer with this story, mostly because it wouldn't feel true. I didn't get into anything that day. After I made the call, I went back to my dorm room and read up on the conference in Copenhagen. Over the next year, I took no further action on the Problem.

What did happen was that Emily and I became friends. By sophomore year we had formed a housing group with four others, and together were assigned a six-person suite in a large, Brutalist dorm near the edge of campus. The building was made entirely of concrete, with mazelike hallways it was rumored had been designed to preempt student riots in the '60s.

I would come home to Emily lying face up on the dorm's thinly carpeted floor, head propped on whatever was at hand— a backpack, a Nalgene—inhaling a novel at alarming speed. Or she'd be folded into her desk chair, scanning Wikipedia and chewing absently on a lock of hair. Her reading sometimes led her to articles about the Problem, which we would discuss with a mixture of hilarity and alarm. The news they contained seemed more than worthy of a riot. Finally, we joked, an excuse to test out the hallways.

For my part, there was never any eureka moment, no single event or piece of news that made me resolve, from then on, to take up the cause. Instead, there was this feeling of slow entrapment: the more I learned about the Problem, the more its importance seemed to balloon, until I could no longer avoid my sense of implication. As if once I'd fixed it in my gaze, my mind could not be made to forget that it was still there, hovering insistently behind whatever else I was occupied with.

In search of an outlet for this muddle of anxiety and obligation, Emily and I joined a campus group that advertised itself as being concerned with the relationship between the Problem and "Development." Each week we'd meet with a professor and a handful of other students, crowding into a small conference room in the university's environmental center. The professor was a kind and disorganized man, with an endowed chair in environmental studies and sociology. As a group, our mandate was never quite defined. We had lengthy, generalized discussions about the challenges the Problem posed for "developing nations," and their need for fair representation at the UN climate negotiations. The professor made allusions to our someday drafting white papers that would support certain delegations from the

Global South, but it was unclear how we would go about doing this, or what expertise we'd bring to bear. Sometimes we would Skype a diplomat from The Gambia, a colleague of our professor who spoke hastily about his delegation's research needs, enumerating various topics we could barely understand. During these calls, I'd glance across the table at Emily, who'd invariably be sunk down in her chair, embarrassed that all we had to offer this learned, harried-seeming man was a room full of bewildered undergrads. The fact that any national delegation would be taking the time to consult our haphazard little group seemed to indicate something very scary about the world's broader level of preparedness.

The last twenty minutes of every meeting would be taken up with the collection of everyone's lunch order for the following week's session. Going around the table, each of us would name our choice of meat or cheese, our preference for either a baguette or a panini. The professor would transcribe each order word for word in his notebook, and at the next session our lunches would be there waiting for us, delivered ahead of time by a local bakery. Emily started calling it The Apocalypse Sandwich Club. After a few months, we'd both stopped attending.

By then, though, there was no turning away. The Problem had insinuated itself into our friendship, lending it the succor and structure of a shared concern. We still mostly talked about other things—our classes, our classmates—but the Problem was like a premise we'd both watched each other internalize. We understood—and understood each other to understand—that it was deadly serious. And despite our gallows humor, we were people who took serious things seriously. It would have ruined our self-concept, I think, to have ignored an apocalypse in the making.

It was in our third year that we finally began to work seriously on the Problem (inadequate phrase, like the whole thing was a term paper, though how else can I convey to you the strangeness, the fixation). Together with some friends, we founded a campaign to divest our university's endowment holdings from the coal industry. Our argument felt obvious: that if the purpose of the institution was to prepare its students for their future, then it could not sustain itself through investment in an industry whose core business model threw that future into jeopardy. At the time, similar campaigns were popping up among students at a half dozen schools on the eastern seaboard, spurred perhaps by the twin failures that had greeted us upon our matriculation: the utter breakdown of that UN summit in Copenhagen, and the defeat of a major carbon cap bill in the U.S. Senate.

We spent hours outside of class collecting petitions in the dorms, circulating sign-on letters among the faculty, and holding rallies outside the administration building. We took the university's insignia—a shield-shaped crest topped with a half-sun encircled by cloud—and turned it into our logo, adding a smokestack to the base of the cloud and giving the sun a frown.

Most of us had little idea what we were doing, but we learned the basics quickly: how to write a press release, how to draft a recruitment plan, how to research and diagram the relative influence of every voting member on the university's board (whose membership, we would learn, included two coal exporters as well as the CEO of a bank with significant investments in the industry). Over time, the campaign became most of what we did outside of schoolwork, and we were constantly thinking up ways to garner it more attention. Around Christmas, we hung a stocking full of coal outside the university president's house and sang theatric carols about the melting North Pole. When spring came around, we inflated giant black balloons in the middle of campus and invited passersby to "pop the carbon bubble." Throughout the year, we would gather in the same classrooms where we'd learned about vertebrate embryology or continental philosophy and hatch our modest plans to avert annihilation. We used the same emails to turn in problem sets as we did to organize rallies. We saw the same people at those rallies that we'd see later at the house parties, cracking beers on someone's back porch. And all of this was comforting in a way—the structural equivalencies could make the campaign feel like it was just another extracurricular. Like it was something we were doing for fun.

Eventually, the campaign generated enough buzz that a campus debate was organized on the question of divestment. We thought this was a good idea, believing, as we still did then, that reasoned debate mattered, and that the Problem might be solved if our side could simply prove its point with enough persuasion. To argue in our favor we invited Bill McKibben, the author and founder of 350.org—at the time probably the country's most famous advocate for action on the Problem. He would square off against Jim Rogers, a former executive at Duke Energy Corporation.

Among his cohort, Rogers was considered rather enlightened in that he was willing to publicly acknowledge the existence of the Problem, though in his opinion divestiture was far too radical a step and did not represent the full complexity of the situation. This was the story he told on the debate stage, where he was invited to the podium to deliver the opening remarks. He wore a red tie and a pocket square and when he spoke it was in a tone of studied moderation, like a weary adult addressing children. Didn't coal, he began, still power a huge swath of our economy?

Did we not have coal to thank for lighting our homes and our schools and our hospitals? Rather than vilifying coal companies, he told us, we should celebrate their hard work, which has made all of our lifestyles possible. Eventually we may need to supplement or even replace coal with renewable energy, but this transition should proceed gradually, with coal and other energy companies at the helm. For now, he said, we are all dependent on coal, and we better start acting like it.

This was something you often saw then in debates about the Problem: normative arguments were met with essentially positive responses. It was a clever trick, really—in a situation so grave, where the changes required were so immense, merely restating the conditions of our predicament was usually enough to deflate debate.

So when McKibben rose to speak, we felt worried. He looked wonky up there, clearing his throat and adjusting his glasses. As soon as he started in, though, it was clear he wasn't falling for the trick. What Mr. Rogers has failed to mention, he said, looking out over the audience, is any of the actual science. What it tells us—what we keep on hearing from the top researchers at the best universities all around the world, universities including this one—is that we have at most a handful of decades to eliminate fossil fuels and replace them with renewables. If we fail to accomplish this, we risk triggering runaway warming so drastic that its consequences will rock the very foundation of our society. Growing seasons will falter, fresh water will grow scarce, the sea will inundate our coastal cities, and the worsening storms and heat waves will make ordinary life increasingly unbearable. To proceed gradually in this moment, at the pace set for us by the coal companies, would be to jeopardize the lives of millions of people, along with many of the places they call home. And this assumes that we can count on the coal companies to even participate in that transition, to drive their own obsolescence out of concern for the common good. But of course, he said, the opposite is happening: the companies from which this university is being encouraged to divest are responsible for one of the most expensive and sustained propaganda campaigns in American history, and have fought every step of the way to prevent the government from passing laws that would eliminate their vast subsidies or force them to internalize the incalculable cost of their externalities. The singular motive of the coal industry—to maximize value for their shareholders—is at direct and obvious odds with maintaining a livable future on this planet; at odds, that is, with the interests of every single student in this room. Certainly, he concluded, we cannot ask them to earn their education at the price of financing an industry that is sabotaging their generation. We cheered at this last line, though it was strange to hear ourselves referred to in the third person, hear our future expounded upon by two men who would never see it.

In his rebuttal, Rogers failed to address any of McKibben's points, and instead spoke in general terms about the importance of things like growth and innovation. If he was embarrassed by this failure, he did nothing to show it, smiling and chuckling as if he'd already won.

As the men went back and forth, our university president sat stiff-necked in the front row, studiously ignoring the band of students who'd come to gauge her reaction. As usual, we'd brought cardboard placards supporting divestment, and we held these at chest height the whole time, though they took only a second to read. After the debate was over, I remember her approaching Rogers as he stepped down from the dais. He seemed to take his defeat gracefully, and the two of them were laughing together. On the other side of the room McKibben stood awkwardly by his chair, collecting his notes. I recall feeling a twinge of unease that night, even as we celebrated the successful event—I was confident our side had won, but I was beginning to doubt if the debate had mattered at all.

It feels important to tell you that there were men (and they were mostly men), far worse than Rogers, men the thought of whom made me shiver with anger. These were the men who not only defended coal but fought tirelessly for its expansion, spending vast amounts of political and actual capital to safeguard the hegemony of fossil fuel, their fervor seeming to increase as the scientific evidence mounted against them. These men spent millions of dollars trying to bury evidence of the Problem, waging a public relations war that cast doubt on research that their own scientists had corroborated—or in some cases even helped to produce. Their behavior seemed so surreal, so unthinkable, that I often struggled to fix it in my head, relying instead on a stepwise ratiocination to remind myself of what we were dealing with.

Imagine a man who was given the opportunity to become extremely wealthy and powerful by pursuing a course of action that had a 1/1,000 chance of decimating society and killing millions of people. A reasonable person would consider this course of action to be profoundly immoral. Even if the action could be done in secret, without anyone else finding out what he'd done, this calculus likely wouldn't change. Now, imagine the same man taking the same course of action, but this time the odds were 1/100. This would surely be deemed unconscionable. 1/10? Pathological. Now imagine that the vast majority of the expert community, people who'd studied their whole lives to be able to understand the potential consequences of this sort of action, concluded that choosing this course would, in fact, have an extremely high likelihood of destabilizing human civilization and killing millions of people. The kind of men I'm talking about are the men who, presented with this final scenario, chose to move forward anyway. At the time, this decision was difficult even to fathom, and so generated both more bafflement and less uproar than you might think.

One of these men was named Scott Pruitt. In 2018—the year I began writing this letter—he was partway through his tenure as head of the federal agency tasked with protecting the environment. What was disorienting about Pruitt was that he fit so neatly into the far more familiar mold of the garden-variety corrupt politician. He was clean-cut and forgettable looking, an avowed Christian who insisted on the constant company of a ten-person armed security detail. He would circle the country in first class, spending ludicrous amounts of taxpayer money attending meetings with the fossil fuel executives he was meant to regulate, hosting carefully curated events with handpicked audiences, giving out shady contracts to former associates. In his first months in office he flew his entire security detail to Italy, where he was photographed alone in front of the Vatican, squinting and smiling into the sun. Back home in Oklahoma, he'd been the proud co-owner of a minor league baseball team, the Oklahoma City RedHawks. It was difficult to comprehend that this was the person responsible for undermining the science that warned of the Problem, dismantling the policies that might have contained it, and pursuing an aggressive course of fossil fuel expansion certain to exacerbate it. It was an evil that was difficult to keep in focus, so the press tended to concentrate more on his pedestrian sleaze, the scandals for which they already had a playbook. Eventually these conventional scandals forced his resignation, though his replacement, a former coal lobbyist—equally bland though less prone to outright graft—would go on to pursue an identical agenda.

The basic problem with Pruitt's Arendtian banality was that it was hard to get anyone to talk about it. Even I found it boring, almost beneath inquiry. Inconveniently, this was a time when great evils could be and often were hidden in the dullest corners of things: the shifting margins of the tax code, the ephemera at the bottom of ingredients lists, the slight changes in the value of a certain type of esoteric derivative. Over dinner, at parties, it was therefore difficult and unpopular to talk about what was going on, which was that Scott Pruitt and his cohort were in the midst of perpetrating what could reasonably be considered a genocide, perhaps the largest of all time.

I want you to understand—though I'm not sure I do, completely—why this was so hard to say; why, even when all the evidence was there before us, it was still difficult to name. I think it had something to do with how much of our ethics were still built on the foundation of intent, reliant on the increasingly irrelevant language of person-to-person causality. Murder, theft, adultery—these were problems for which we had a script. Even typical human rights violations, perpetrated directly by a government on its people, were usually legible. But we'd been thrust into a world where the most influential moral currencies—data, carbon, capital—were circulating far above our heads, through networks of impossible speed and complexity. Our technological innovations had outpaced our moral innovations in both ingenuity and popular uptake, and so we were stuck wielding localized ethics in the face of globalized problems. Loosed from the bounds of intuitive causality, then, sociopathy was easier to mask, harder to question. And though the best science had made clear how Pruitt's actions could irrevocably compromise global food and water security, starving and then killing millions of mostly impoverished people, and forcing millions more into massive migrations through increasingly volatile geopolitical terrain, it still sounded uncouth, even a little ridiculous, to spell this out in conversation.

Here is a hope: that in your time, they've developed an ethics lucid enough to condemn Pruitt without him having to personally drown ten million people (his security detail dragging them screaming into the ocean, in Tuvalu, in Bangladesh, in Far Rockaway). And another: that his name lives on, utterly ignoble, that it's been turned into some kind of epithet.

The former CBS news anchor Dan Rather once proposed carving the names of the deniers into a giant monument on the coast of Miami, a reminder for posterity. Another method might be to mark the gravestones of those they killed. For example, it was once common practice to make special note on a headstone if someone had died at sea. I am imagining the graves of all those people drowned in Sandy and Maria, the families burnt alive in their cars in Paradise, the farmers who took their own lives after another failed harvest—an entire cemetery carved with the words "Death by Pruitt." Ugly thought, I know. As Jim Rogers would say, correctly and irrelevantly, none of this "reflects the full complexity of the situation."

There is another part of me that wants sincerely to establish grounds for empathy with Pruitt. I can't even begin to access the mind-set that's allowed him to do what he's done. And as with art and locked drawers, inaccessibility can impel curiosity. I half suspect that what I'm actually doing here is trying to bait a libel lawsuit, the only strategy I can think of for placing us in conversation.

This is a trap, I know: the assumption that we are both just individuals, that a conversation would even be possible in the prosaic sense of the word. Of course Pruitt *isn't* just an individual. He is also a proxy, an embodied stand-in for a whole convoluted web of wellheads and stock prices and anxious investors.

Here is a serious question: How can you talk across that kind of gap, the merely corporeal to the vastly incorporated? What language would you speak?

This question would come to haunt our divestment campaign. After months of advocacy, we were finally granted a meeting with the president in her office, a wainscoted room in the university's administration building. When we walked in, she greeted us warmly, like we were already in agreement. We were invited to sit down at a long oval table, and were each given a tumbler of water on a coaster emblazoned with the university's logo. The suns on the coasters were smiling.

"Let me first begin," the president said, "by saying how truly inspiring all your work has been, and how robustly you've contributed to our university's conversation over the past year." However, she went on, the university board had a fiduciary duty they could not breach, and that was to maximize investment returns on the endowment. Divesting from coal would violate that duty and instrumentalize the endowment for political ends. It simply couldn't be done.

Though we knew then that we'd lost, we tried one last time

to make our case, almost pleading. The coal industry was setting us on a path to catastrophe, we told her, and it was our generation that was going to bear the brunt. Air pollution from coal was already killing hundreds of thousands of people every year, and that was without any tabulation of the deaths caused by worsening storms and droughts. It was our moral obligation to erode the industry's social license, to withdraw the imprimatur of our investments.

When we had finished, the president looked at us and laced her fingers. "I do not agree," she said, "that this is primarily an ethical issue." There was little more to discuss. In the end, it seemed, we'd been speaking the wrong language.

Soon after, a month before my graduation in 2013, the university released its official statement on the matter: "After serious, thoughtful, and robust discussion . . ."

———

A FEW YEARS LATER, I was driving home from a conference in Atlanta—I'd participated in a panel on "equitable carbon pricing"—when my car ran out of gas on a highway entrance ramp. I got out and pushed it onto the median, then waited there with my thumb out. Other cars shot past, their speed suddenly palpable from a few feet away. The grass of the median had little burrs in it, and they kept sticking to the ankles of my jeans.

Eventually the driver of a pickup truck offered me a lift to the nearest gas station, a couple miles down the road. When we got there the station looked empty. Red numbers flashed on the price display, and a pavilion shaded the pumps. Nearby, a giant combine grazed through a wheat field, throwing up fountains of chaff.

I stepped inside the little store and looked around for a container, deciding on a gallon jug of water, which I emptied out in the parking lot. The water cut the dust from the chaff and left a stain on the asphalt of slightly blacker black, the color I loosely associated with gasoline. But when I stuck the nozzle into the neck of the jug, what came out was yellow and half translucent, like corn syrup.

I stepped to the curb and stuck out my thumb again, but this time no one stopped. After half an hour I decided to walk back to the entrance ramp, hefting the gallon. The shoulder of the road was strewn with gravel and shreds of rubber, like the molt of a large black snake. I felt embarrassed carrying the jug, like I was revealing some compromising secret about myself to all the passing cars, whose passengers I variously imagined to be you or me or Pruitt himself. How I'd wasted a gallon of water to make way for a gallon of gas. How I hadn't even known its color till it was jostling against my thigh.

After what seemed like more than an hour, I got back to the car, still at rest in the weeds. I used a pen to push aside the metal flap and poured the gas from the water jug into the tank. That was all it took. The car revved to life, the gas got me home.

Here was the gotcha moment. *Aha!* say the Pruitts in my head, gloating now. *Without oil you wouldn't have slept in your bed that night! You wouldn't have been able to even give your little talk in the first place!*

Usually these inner arguments aren't hard to refute. Of course campaigning to wean ourselves off fossil fuels does not eliminate the fact of our current dependence. Of course the future we want does not erase the present we have. (Though many

environmentalists fall into exactly this consumer's trap, focusing more on eliminating hypocrisy from their purchases than on
building the political power necessary to overhaul energy systems and obviate precisely these kinds of impossible decisions.)
And I know that these lines of reasoning elide the vast difference
in our relative power, how the Pruitts shape the energy landscape the rest of us are forced to occupy; how our choices are
nested inside of theirs.

But still, but still. I want you to know that every time I pulled
the trigger on the pump, a small dose of loathing came up with
the gas—loathing for them, loathing for myself—and though I
knew it wasn't right, I couldn't always stop its spread. Sometimes it'd seep through me, unseen but black as bile.

———

DURING COLLEGE, in the years when I was consumed by the divestment campaign, I enrolled in a geology course called Earth:
Evolution of a Habitable Planet. The professor, a temperate man
with glasses and a goatee, liked to use an old-fashioned slide
projector, on which he would place translucent plastic images of
granite or gneiss, painting the wall of the classroom in various
shades of stone. One of his favorite things to repeat during lectures was that carbon was Earth's temperature dial, and whenever he said this he would twist his fingers in the air, like he was
turning up the volume on a car radio.

For the final exam, we had to memorize all the geologic eras
and epochs in sequence, as well as their start and end dates. They
had fabulous, arcane sounding names: Ordovician, Silurian, Mis-
sissippian, Permian. The current era, he told us, the one we were
in now, was called the Anthropocene, a term invented to replace
the Holocene, the most recent period, which for decades the geo-

logical community had naively assumed was still under way. The Anthropocene, in contrast, was marked by the advent of humanity as a geologic force, by the fact that a biological agent was now manipulating the planet's circulatory systems at their most fundamental level, diverting water and carbon on a scale usually accomplished only over millennia.

This is old news by now, and the term has since been contested by various alternatives. (In light of the Pruitts, some scholars have suggested the term Capitalocene or Plantationocene, a reminder that not all Homo sapiens share equal responsibility for the Problem.) Still, at the time, the word "Anthropocene" knocked the wind out of me. Just the fact that we could be living in it and not know, or know, but not fully register.

The course was held inside the geology department building, near the center of campus. Through its windows you could see the roof of the administration building, where, eventually, the decision would be made to maintain the university's investments in the industries most responsible for this sudden and drastic change in geology curricula. The two buildings looked very similar—red brick and white trim, done in the Georgian style. But crossing the quad between them I would be struck with a kind of vertigo, like I was walking back into the Holocene, passing from one reality into another, impossibly similar yet completely distinct. On the way I would sometimes stop at a student café, where this disjuncture could be soothed through the purchase of expensive, organically sourced salads, which you could then shit out into toilets that had water-saving adjustable flush mechanisms. The handles were green with a little insignia of our habitable planet on the end. You pulled left for number one and right for number two, and either way it was all quickly whisked away.

When this dissonance became too great, I went looking for

alternative framings, something to offset the dissociative trance that seemed at times to animate the entire university, inhabiting even the geology department itself, which every year—after instructing them in detail on the likely consequences of continued fossil fuel dependence—sent many of its graduates to work for companies where their expertise would help extract precisely those fuels from chthonic shale deposits and deep-sea basins newly exposed by melting ice.

In this respect, history courses were particularly useful. Reading the textbooks, I would keep mental tabs on all the moments in the past when the world had seemed to end. The year 70,000 BCE, when the eruption of a Sumatran super volcano threw the whole planet into ashy darkness for years, decimating the already small population of Homo sapiens that had lately sprung up in East Africa. (Some studies have shown that the entire species was reduced to hundreds, even just dozens of breeding pairs, small naked bands picking withered berries under a sunless sky.) Or the thirteenth century, when the Black Death killed as many as two out of every three people in Europe, leaving the survivors spiraling into a state of religious panic, which in turn precipitated the prophylactic slaughter of entire villages of Jews. Or the sixteenth, when the Aztecs saw their centuries-old civilization—an empire that built vast pyramids, that was the first in the world to institute mandatory universal education—crumble in a few decades at the hands of Spanish imperialism. The list went on: the slave trade, the AIDS epidemic, indigenous genocide—with the right historical comparison, I could convince myself that the Problem was nothing new, and this tempered the feeling of disorientation, replacing it with a low-level weariness that made everything feel predictable, and therefore, somehow,

a little more tolerable. Human history, it seemed, was a history
of cataclysm. The Problem was different in scale but not in kind,
a continuity rather than a disruption (or rather, a continuity in
disruption).

But this is precisely why the Problem feels so unbearable. It's the
utter familiarity, the numbing repetition. The sense of bearing
witness to the same worn-out travesty. Wherein profit is cen-
tralized and risk is distributed. Wherein the powerful sacrifice
the vulnerable to the god of growth, then feign deafness when
they cry out in protest. They are too poor or too Black or too far
away to hear. We do not know them, and their plight is not our
concern.

In this sense, the nature of the Problem merely restates the
moral equation of colonialism: Once again, the Pruitts of the
world seek to enrich Western corporate interests by jeopardizing
the lives and livelihoods of millions of poor people, most of them
in the Global South. Except this time the pattern has grown so
obdurate and powerful and maladaptive that it threatens to upend its own premises, shaving tens of points off the global GDP
in whose name it is rationalized.

Emily was also enrolled in the geology course, and we'd often
study together in an empty classroom, memorizing flashcards
with the names of distant eras and obscure species of rock. It
was really bad luck, she liked to point out, that the Problem was
causing temperatures to rise instead of fall. Imagine the coun-
terfactual. Chicago, Moscow, and Beijing snowed in for months,
frostbite warnings sent out as emergency alerts, ice on the highways and clogging the ports. Meanwhile, from Laos to Gabon,

the equatorial latitudes would be getting their first taste of sweater weather. We probably would've tackled this thing forty years ago, she'd say, joking, but also not at all.

————

IN RHODE ISLAND, it would snow only a handful of times each winter. When it did we took to the streets, Emily and I and the rest of our friends, wandering for hours down the whitened sidewalks. I remember one blizzard that pummeled the city for a full day, drifting up the sides of houses and coating the telephone wires in ice. It was still going when dusk fell, and you could see snow falling fast under the streetlights.

A group of us walked down to a park by the harbor, moving with our heads bent low, our hoods cinched against the gale. From the wharf that fronted one end of the park, we watched the snow disappear into the water, falling in sheet after sheet into its face, a monstrous force with no cumulative effect. Looking out, you couldn't see the far shore at all, just snow and water receding into gloom, as if somewhere out there the world just came to an end.

On the way back we saw other bands of dark figures, anonymous in their hats and heavy coats, loping through the thin light cast by the few open stores. The swirling dark seemed to loosen something, lend a strange kind of license, and as we passed them everyone would start tossing snowballs at random, howling beneath the wind, before stumbling off again into the storm. The strength of the wind made it impossible to talk, whipping away our shouts as soon as we opened our mouths. In the darkness, the snow looked like graphite smudged across the houses, through the branches of the trees. It was as if the apocalypse had come early, with all its chaos and thrill, and we were no longer students at all but survivors, ducking around the corners of an obliterated city, inaudible and almost invisible.

Though there was the opposite feeling too, sometimes, a prelapsarian nostalgia for that period—just a century ago, well within the lifespan of that venerable building where the president would make clear to us her position on coal—when snows like this were the norm, when winter was defined by them. Back then, I imagined, I might have stepped outside and let myself fall into the softness of a bank, the weight of my body pressing slowly into the snow, the flakes drifting downward to cover my face. I imagined my digits growing numb, my brain losing track of them in sequence. And along with the warmth there was always a mild disappointment when I finally stepped inside and felt my fingers coming back, the edges of myself thawing into place. I think now that this longing was one response to learning about the Anthropocene. Because a part of me wanted to believe that we were still subordinate to the world; that the snow might erase us overnight.

Instead, the opposite was happening: we were supplanting the snow, or at least implicating ourselves in its fall. After his debate with Jim Rogers, I picked up Bill McKibben's book *The End of Nature*, one of the first ever written on the Problem. "By the end of nature," he wrote, "I do not mean the end of the world. The rain will still fall and the sun shine, though differently than before. When I say 'nature,' I mean a certain set of human ideas about the world and our place in it."

McKibben had used the phrase figuratively, but there are those now who take the end of nature very literally, who in fact believe it has already happened. The theory—ascendant among certain schools of metaphysicians—is that we are all living in a sort of digital simulation. Given current computing trends, these philosophers argue, it seems not at all implausible that one day, possibly soon, we will develop the processing power to create giant virtual worlds inhabited by avatars who can learn, reflect, reproduce, wage war, and write speculative dissertations in philosophy. If this assumption is reasonable, then who's to say it hasn't already happened—that we are in fact not the creators but the inhabitants of this virtual reality, and that what we take to be our psyches are merely flickering patterns on a series of semiconductors in an outer world eye hath not seen. Or perhaps, they say, we are both avatars *and* creators, and each successive generation of conscious life eventually takes the Rubicon step of creating a smaller reality within itself, leaving the virtual denizens to scan their palimpsest for hints of the reality beyond. Which would mean that there is actually a whole nested series of virtual worlds, leading all the way up to an original creator, a Russian doll so old and massive that we can think of it only in metaphor, and even then imprecisely. The seduction of the theory is that things like the Problem, which appear cataclysmic on their face, turn instead into relatively minor bits of algorithm determinant of certain patterns of events in a virtual world whose plot its creators—perhaps trying to distract themselves from the capricious and invisible whims of the Russian doll above *them*— have decided to thicken a little.

By 2012, the year we started our divestment campaign, some version of this theory had taken root among the twin clerisies of finance and tech. Even if nature hadn't already ended, the more ardent disrupters seemed intent on ending it. Their ideas would catch my eye in the university library's periodical room, where they'd begun to furnish headlines in some of the popular science magazines. Freezing our brains and uploading their contents to the cloud; handing the reins of the economy over to artificial intelligence; living forever in a customized virtual paradise. The articles were shocking to read, less for the predictions they laid out—which often seemed hazy and implausible—than for what they indicated about the direction of debate. Because while we were campaigning for the preservation of a habitable planet, here were people who appeared honestly to want to obviate it, to finally unpeel humanity from any dependence on or relation to the physical environment.

The most influential futurists of the day (I'm thinking especially of Yuval Noah Harari) had latched on to the whole scenario as a sort of neo-Noahide race: either we would build our digital ark or drown in a flood of our own making. Though given the "we" they were talking about—the wealthy few for whom this scenario could ever seem even remotely feasible enough to provoke real suspense—I pictured instead a wood-inlaid Cessna, idling on a private runaway as the water took the tarmac.

Barring a flight to Mars or the Cloud, however, the rich may fall back on earthlier escapes. Even as they proselytize about the coming digital utopia, some of these same Silicon Valley luminaries and hedge fund billionaires are buying up property that might withstand a civilizational unraveling. The astronomically wealthy have started to snatch up large plots of farmland on New Zealand's South Island, abundant in fresh water and isolated from major population zones. The merely very wealthy are beginning to purchase shares in luxury bunker complexes, some

of them built inside abandoned missile silos in places like Topeka, where a cottage industry of boutique preppers has outfitted them with hydroponic gardens and oxygenated tilapia tanks,
together with giant flat-screen TVs displaying beautiful images
of places like the Bahamas or the Rockies—white sand beaches
and snowy peaks which, in the event the bunkers become necessary, may no longer exist, and will anyway be inaccessible
beneath several dozen meters of Kansas clay.

But the hedging runs deeper than this. Even in a collapse—
especially in a collapse—there are bets to be made, profits to
be turned. There are stories of investors buying up short-term
housing in hurricane-prone cities; of agribusiness gobbling up
family farms that have been broken by drought; of oil companies
purchasing previously unusable patches of the Arctic; of traders
going long on water rights and desalination plants and medica-
tions for malaria.

I want to be able to accept the received wisdom here: that
short-termism has simply driven these people mad, winnowed
their ethics down to the quarter, engendering a slavish psychosis that compels them, even as the ship sinks, to hoard more
and more gold in their cabins. But there is something about their
psychosis—so obdurate, so *eager*—that raises a further question, one that I can scarcely entertain. Because what if all the
defensive stockpiling, the relentless investment in the means of
destruction—what if all of it is a sign of something more sinis-
ter? That the Pruitts aren't blind, but entirely calculated. That
the Problem is not a form of fallout, but rather its own kind of
weapon.

I do not subscribe to any grand conspiracy on this. I do not
think the superrich host secret meetings, or monthly conference

calls, or even talk to one another much at all about the Problem. But I'll admit it has crossed my mind that perhaps it's crossed theirs that on an overpopulated planet where human labor is obsolescing, the consequences of the Problem might not be so unattractive. Provided they could amass a fortune large enough to survive it, a mass die-off could prove convenient, even profitable.

Sometimes this thought feels so intolerable that I catch myself half hoping we really are living in a simulation. That the reason things seem so surreally bad is precisely because they aren't real at all. Even in college, when I first encountered the idea, I could sense its guilty appeal.

I'd enrolled in a philosophy course taught by a professor who was compelled by the theory. Though she did not subscribe whole-sale, she'd decided early on to hedge her bets. If we were indeed living in a simulation, she told us, then perhaps our creators were using our reality not just for entertainment, but to model the results of certain sets of historical, even physical, variables on the system as a whole. As with any computer model, she was banking on it one day being run again, and she strove to be so outlandishly interesting that the creators wouldn't be able to re-sist keeping her avatar around for the next iteration.

I'd often see her shuffling across campus, a tiny woman in her seventies wearing a golden turban and multiple dresses made of some sort of voile. In lecture she wore a giant clock face on a chain around her neck and would occasionally heft it upside down in front of her to check the time. Surely, I reasoned, she was a top contender for virtual resurrection.

On one occasion she called me into her office and, after a brief discussion of a term paper I was writing, put it aside and looked at me over her bifocals. "So tell me about this environmentalism," she said. "I've never understood why I should care about one species of owl versus the next." I couldn't produce a satisfying answer, probably because in that moment I wanted so badly to inhabit her point of view. How hilarious, how arbitrary owls seemed in the light of the simulacrum.

———

I SHOULD CLARIFY that I am not an environmentalist, though people often assume that I am. Once I saw the director Darren Aronofsky speak at a gala to raise money to save the environment (the unstated assumption being that with enough money its protection could be purchased). Between the canapés and the main, he took the stage wearing a trim suit and glasses. "I'm Darren Aronofsky," he told the crowd, "and I'm a fucking environmentalist." I clapped along with everyone, though this was precisely why I hated the word. I could hear how it needed that supplementary "fucking," how this was the only way it could still be made to sound unapologetic.

I traced my misgivings back to the origin of the word "environment," which derived from the humble French preposition *environ*, meaning, simply, "around." It didn't find its way into the English lexicon until the mid-nineteenth century, when it was defined as "the aggregate conditions in which a person or things lives," a concept so vague as to just barely distinguish itself from the word "everything." But by the late twentieth century, the word "environment" had been cordoned off, dislodged from all the other things we cared about, sometimes even crowned with a capital E. There are scholars who claim that this fracture coincided roughly with the invention of the steam engine. Others trace it back much further, to the advent of the Abrahamic faiths, the shattering of the animist idols. Either way, by the time I was old enough to use it, the word seemed to invoke a species of paternalistic love, a de facto separation. The environment was a whale in danger, a beautiful but distant forest. Everyone wanted to protect the environment, but few engaged with it regularly.

The mother of a friend of mine liked to tell a story about her early days as an environmental lawyer. One of her first assignments was to help conduct an environmental impact assessment for a proposed expansion of an upstate regional airport. She drove up to meet with the head of the county transportation department to walk him through the process. "You don't have to bother," he said, when she told him what she was there for. "There's no environment up here, it's just a bunch of trees and scrub." This was how the process worked on the ground.

Once it had been set apart on its pedestal, the environment warranted its own belief system and so took its place among the other faiths: fascism, veganism, Catholicism, etc. Transformed by its doctrinal suffix, the environment was now one alternative among many, a choice at the buffet. And seeing what had happened, a whole generation of well-meaning adherents scrambled to (often literally) sell this particular choice back to us, emblazoning a liturgy of exhortations and admonishments on bumper stickers and T-shirts and newspaper columns and picket signs and legal complaints and campaign speeches. Here were Save the Whales and Go Green; here was a single polar bear staring out at you from the side of a tote bag, provoking a twinge too feeble to bridge the vast semiotic distance that separated you from it, standing there on its fabric floe, adrift off the arm of a stranger.

As it was, I struggled to feel passionate about environmentalism—this cheapening of a God into a religion, in need of

legions of proselytizers just to stay relevant. I wanted to erode
the whole word, to eliminate its usage and smuggle whatever
feelings it attempted to transmit back into the bedrock of language. I wanted to bury it so deep in our vocabulary that it'd get
reabsorbed, like how a body buried in earth becomes it.

For a time, I went in search of other words, subscribing as I did
to Joan Didion's mildly self-serving maxim: "that the ability to
think for oneself depends upon one's mastery of the language."
Perhaps, I thought, the Problem was born of bad thinking at
scale, and might be remedied if we could simply find the right
words. This line of inquiry quickly veered into something more
incantatory, a search for a spell I could cast that would engender
in people a new and, at last, fully motivating sense of what was
at stake. Sometimes I would try "Earth," but it had been wrung
dry, already in use for decades as a shibboleth for the kind of
consumerist environmentalism that promised salvation for the
price of a compact fluorescent lightbulb. Whenever I used the
word, I would picture this beach ball I had growing up that was
printed to look like a globe, which sat for years partially deflated
in the corner of our basement. *Earth.* It held an unmistakable
gloom, like you were recalling a missed opportunity. "Planet,"
too, had been played out, in the infographics of middle school
science textbooks and the basso profundo of Neil deGrasse Ty-
son. If anything, "planet" rang of an oblivious optimism—"planet"
was the blue marble in the shuttle window, an amazing system
we were just beginning to understand. Its typical inflections
gave no indication that something had gone seriously wrong,
that the Problem had placed a frightening asterisk over the
whole ordeal, suspended it above our heads like the sword of

Damocles. For a time, "world" seemed like the best option. To my ear, it was tender and unpretentious; it captured the thin atmosphere of collective memory surrounding our beloved little rock. World, as in the whole. World, as in the end of. Still, when people asked why we worked on divestment, I could never bring myself to say what I felt: that we were trying in all seriousness to save the world. I couldn't imagine anything more clichéd.

At some point I read a seminal essay on the Problem by Zadie Smith. "There is the scientific and ideological language for what is happening to the weather," she wrote, "but there are hardly any intimate words. Is that surprising? People in mourning tend to use euphemism; likewise the guilty and ashamed." After this, my search grew less determined. Most likely I'd been looking for something that wasn't there.

———

TO DEMONSTRATE TO US the consequences of adjusting Earth's temperature dial, our geology professor once showed us a website where you could visualize the rise and fall of the oceans. We spent the class dragging the cursor back and forth through geologic time, watching the green continents submerge and change shape, reappearing again like the backs of crocodiles. As you scrolled from the Ordovician to the Eocene, a numerical display in the top right corner toggled through the attendant parts per million of carbon in the atmosphere. If you scrolled into the future, the map displayed various predictive scenarios of twenty-first-century sea level rise.

After we'd played around for a while—fingers grazing lightly across touchpads, continents contorting through time—the professor zoomed in on our city, to see what our own future had in store. The flood we witnessed was nothing biblical, nothing cleansing. The ocean simply crept upward, millimeters each year, a geological sprint and a biographical crawl. We watched until the pixelated blue had dissolved the coastline, taking bites here and there, like an insect nibbling the edges of the map. We watched until the water began inching up the hill toward our classroom. Then the lecture ended, and we walked out into the sunshine, toward whatever class we had coming next.

Our professor used to describe the Anthropocene as a "slow-motion emergency," though he somehow avoided this ever sounding dramatic, inserting it instead like a minor footnote, a fact too obvious to be belabored. This was a very difficult concept to fit into a story. "Slow-motion emergency" evoked no ultimate triumph or tragic catharsis or even so much as a discernible ending, just a long slide into loss and uncertainty, a literal and littoral deterioration. The baseline was shifting too quickly for policy and too slowly for narrative (perhaps unsurprising, since the latter often impelled the former). And so, outside of class, Emily and I and the rest of the campaign were left stranded with a jumble of facts, trying to piece them into some arrangement that would at last induce in the administration the same sense of urgency which we would feel for a moment in lecture and then lose, hours later, its sharpness dissipating into long, grassy quads, the stately permanence of the brick.

The Pruitts, by contrast, had excellent stories on their side, true blockbusters of progress and redemption. We knew their stories were good because we'd read them so many times—in the editorial pages of *The Wall Street Journal*, between the lines of Jim Rogers's speech. Their narratives were easy, but more than that they were validating. In them, America got to achieve global dominance not despite but *because* of our fossil-fueled lifestyle. Profits would soar, freedom would reign, and these statements would always be redundant. No one could tell you what car to buy, or how to water your lawn, or where to throw your trash. The most patriotic thing you could do was whatever the fuck you wanted. And if this ever backfired, God would step in to protect those faithful who had, despite all warnings to the contrary, continued to ignore the science, trusting completely in His grace.

This is all a paraphrase, of course, and maybe it's too parodic. There were many justifying stories far subtler than this: that fossil fuels were getting cleaner, that they could not be replaced, that the consequences of trying to do so would be worse than the warming itself. These stories appealed to other gods, invoking the omnipotence of "the market" and "the economy" (or, in the case of our university president, "fiduciary duty").

What scared me was not the content of the myths, but the fact that there could be stories so compelling and gratifying and widespread and long running that in order to maintain their integrity, millions of people would be willing to sacrifice the world they purported to describe.

The more involved I got in the campaign, the more desperately I sought out other stories, stories that might make sense of the Problem—not what it was, or how to "solve" it, but how it felt to be alive in its context. I looked mainly in the fiction sections of bookstores, the aisles I gravitated toward naturally, which always seemed to promise a meaning beyond bare knowledge. But with very few exceptions (Ben Lerner's *10:04*, especially), all I found were glancing references, some one-off plot devices (e.g., the sea wall that appears around Manhattan in the final pages of Jennifer Egan's *A Visit from the Goon Squad*). No body of work that captured the frightening scale of the Problem. Its absence made me feel lonely.

I did eventually find the next best thing, a book that attempted at least to explain this lacuna. In *The Great Derangement*, Amitav Ghosh approached the Problem not through shifting baselines but through the cataclysms that punctuated them—the violent hurricanes, the calving of icebergs bigger than U.S. states. "The extreme nature of today's climate events makes them peculiarly resistant to contemporary modes of thinking and imagining. This is particularly true of serious literary fiction: hundred-year storms and freakish tornadoes simply feel too improbable for the modern novel."

I didn't disagree with him, but after reading Ghosh my paralysis deepened. Because if it was too boring to write about the trends, and too artless to write about the events, then what *were* we supposed to do? (Resign ourselves to speechlessness? Let the Pruitts fill in the gap?)

————

WHY NOT JUST REVOLT? I can picture you asking. Why didn't our movement simply burn the instruments of your preclusion, dismantle those machines that every day devoured more of your future, like some cheap and unprocessed ore? The plants and the valves and the pipelines; the tractors the size of houses clawing the dirt of their watery pits.

I admit that I thought about it. How it would feel to slip a cartridge of explosive under a cold length of pipe. Crouching to the ground on some windswept prairie and pressing my head to the metal, reaching with one arm under its belly to feel for a joint. How afterward we would watch the flames from such a distance that they would make no sound, just a silent orange flag unfurling again and again above the wheat, alone and oddly specific.

There were reasons we refrained, obviously, though these have more to do with strategy than justification. Because of course it would have been justified. At the very least justifiable. We knew that, over the next century, the continued viability of these machines would guarantee the loss of millions of lives. Of what import was "shareholder value," was "property," was any amount of charred and twisted metal when measured against the right of human beings to live?

But it wasn't a question of what was right or defensible. It was a question of what would work. From our vantage, any attempt at widespread demolition seemed likely not only to fail, but to backfire. What better ammunition to hand the Pruitts, who would gleefully label us terrorists and then continue, with newfound moral license, to lay waste to the world ("terrorist" being a word invoked most often to defend a status quo, even and perhaps especially when that status quo entails genocide). And even if there existed a large enough political constituency prepared to support the destruction of fossil fuel infrastructure, the state retains its monopoly on force. The dismantling would never be allowed to reach a scale where its drawdown impacts would rival those achievable through policy.

I am almost sure this is true, but still I worry. Because no one knows what the coming decades will bring. If the Problem gets bad enough, which ideas will be vindicated, which decisions ex-humed? Which desperate acts will look reasonable in the ruth-less glare of hindsight? (Which is to say, under your scrutiny.)

———

IN 2017—THE YEAR the eponymic Pruitt was appointed—a hurri-cane named Maria killed three thousand people on the island of Puerto Rico, a death toll ten times that of Sandy. Most of the people killed were poor. Many of them were not white. They were American, technically, though Puerto Rico remains a ter-ritory, not a state, and for this reason none of those killed had been able to vote in the previous year's presidential election, in which Donald Trump had been installed in the White House.

After the storm had passed, Trump made a perfunctory ap-pearance at an emergency supply center in San Juan, tossing rolls of paper towel theatrically into the crowd, like he was shoot-ing freebies into the bleachers at a basketball game. "They had these beautiful, soft towels. Very good towels," he said. "I was having fun, they were having fun. They said, 'Throw 'em to me! Throw 'em to me, Mr. President.'"

Later he would try to erase the postmortem, claiming that the number three thousand had been deliberately concocted to make him look bad. The tally would eventually be confirmed or exceeded by various independent surveys, but by then, presum-ably, nearly half of those Americans who *had* been allowed to vote believed him: that the three thousand people they'd heard about had never died, that they had never even existed.

Here was the other side of the Pruitts' story, the photo neg-ative of that myth of freedom through consumption. In Trump's attempt to expunge the deaths, you could see the outline of the logic that allowed them to purport concern for the plight of Puerto Ricans while refuting the science connecting it to the Problem and pursuing policies that would make the Problem worse. Undergirding all of it, unexamined and cocooned in rationalization, was this essential premise: that there was a certain kind of death—foreign, catastrophic, statistically rendered—that could be an acceptable and ignorable by-product at the end of an important supply chain. That—and this is where it really boiled down—some lives were just worth less than others.

I should pause here to impart my own premise, probably the most important and least original thing I'll share with you in this letter: that all lives are of equal and inherent value. That evil is the slow and always-deniable erosion of this fundamental understanding.

This isn't a narrative; it is a plain, prosaic truth. It goes without saying, which is also to say that it's easily drowned out.

And it has been—consistently, historically—buried under the suffocating weight of the Pruitts' canonical story. From slavery to structural adjustment, the story has allowed them to feed themselves off the bodies of others, a form of cannibalism.

In *Between the World and Me,* Ta-Nehisi Coates refers to this story as the Dream. The book is a letter written to his teenage son on the subject of being Black in America. The Dream, Coates explains to him, is "the lie of innocence," the mask obscuring the habit of plunder. "It is perfect houses with nice lawns. It is Memorial Day cookouts, block associations, and driveways. The Dream is treehouses and the Cub Scouts. The Dream smells like peppermint but tastes like strawberry shortcake." It is "the same habit that endangers the planet, the same habit that sees our bodies stowed away in prisons and ghettos."

The Dream can kill three thousand people in the name of carbonized growth, and erase their deaths in the shroud of whiteness, and to the mourners offer only paper towels—"beautiful, soft" paper towels—those quintessential symbols of guiltless consumption and suburban prosperity.

The Dream is what put the Pruitts in the White House. The Dream is why they lie about the Problem, for any concession to its severity would compromise the delusion from which they derive their power. And it is important to remember that there is a history to all this, that the global fossil fuel industry may be new, but the story that justifies its hegemony is as ancient and buried as the coal it digs up from the ground.

The only difference now is that the Dream is starting to backfire. It has finally produced a Problem that does not obey its laws or respect its heavily fortified borders. Though it often strikes poor people first and worst, the Problem will never confine its violence to them alone. Ultimately, even the Pruitts won't be exempt. Coates again: "Once the Dream's parameters were caged by technology and by the limits of horsepower and wind. But the Dreamers have improved themselves, and the damming of seas for voltage, the extraction of coal, the transmuting of oil into food, have enabled an expansion in plunder with no known precedent. And this revolution has freed the Dreamers to plunder not just the bodies of humans but the body of the Earth itself. . . . It was the cotton that passed through our chained hands that inaugurated this age. It is the flight from us that sent them sprawling into the subdivided woods. And the methods of transport through these new subdivisions, across the sprawl, is the automobile, the noose around the neck of the earth, and ultimately, the Dreamers themselves."

Trying now to write my own letter, it's become clear to me how unprepared I am. I have not grown up enduring the violence of structural racism. I have not had to contend with any immediate and pervasive threat to my body, and the love of my parents did not have to come freighted with this fear. For my own father—like me: white, middle class, college educated, and male—raising children did not require much grappling with questions of injustice and risk. Though he'd chosen to have me, he owed me no account of why. This is what is meant by privilege: that the world had been built with me in mind and he could enlist me in it with confidence.

But now the world has begun to imply an existential threat to even the most sheltered children—the threat of fire and flood, of a less and less habitable planet—and so I'll admit there are times when I harbor doubt, when I question whether or not I should have you.

This questioning is far from new. Millions of people before me have had to consider the prospect of a child in a context made hostile by the Dream. They've done this from plantations and refugee camps, reservations and war zones—places far more devastating and dangerous than anywhere I've ever been; places where a world was ending or had ended. And though it may have felt hopeless and reckless and futile, sometimes from out of this grappling there came a child. I still cannot fully understand the depth of love it took to do this, to loft a tiny salvo of life into a death as wide as the sky. It amazes me. So too the other choice, the one we hear less about. Those families who chose to truncate their tree rather than see its branches hacked at and burned.

In the moments I harbor doubt, both decisions strike me as impossible, impossibly brave. Though framing it as a decision ignores the many people throughout history for whom procreation was never a choice. The women who had their reproductive freedom seized by men. The families who needed children for labor, who would have starved without them.

In comparison, I have real options, which implies both a massive privilege and an unavoidable question: Should I have you, and risk putting you in harm's way? Or should I not, and prevent there ever being a "you" to be harmed? I want to name this tension. Not to say that my ethical doubts have disappeared, or that I've found the "right" answer. But just to acknowledge that, for the purposes of this letter, my heart has tabled the question. I am not asking here whether I should have you, but what I owe you if I do.

As I write this, there are billions of people who have just had children, or who are planning to, or who have yet to decide, and all of us are scrambling to figure out what we can possibly tell you in the shadow of the Problem.

I'm imagining now all the letters being composed simultaneously to this one, all the "you's" they must address. Letters of panic and love and counsel and warning. Letters the bulk of which are probably not written but drafted wordlessly in the backs of minds. I imagine them spilling out from the pens of my generation, piling up in the hands of yours: an unsolicited, worldwide bequest of concern.

The fantasy is that these letters will resolve something, and perhaps they will for us, perhaps they will unburden us of some of our questions. But for you, I know, these questions will only return, likely with greater urgency. At some point later this century you may be faced with writing a similar letter, and assuming the Problem continues apace, that letter will be even harder to write than this one. Though I've tried, I cannot begin to imagine what it will contain, or how you will go about writing it. All I can think to say here is that I'm sorry. My love now comes freighted with this fear. I can't send you one without the other.

———

SOMETIME AFTER WE'D STARTED the divestment campaign, but before we learned we had lost, Emily and I had a conversation about children. As I recall, it wasn't a discussion of whether that time but of when. To be honest, I remember her saying, I think I should really be having them as soon as possible. That way they still have a chance of growing old without ever having to see 2100.

Over the course of our long, commiserative discussions of the Problem, 2100 had taken on an arbitrary significance for us. In our heads, we'd designated it as the year when the social fabric might begin to unravel in earnest, the dividing line between two very different centuries.

I remember how, when she said it, I began to picture you at a nursing home somewhere in 2099, lying in bed with the window open in January, watching the parched grass bake in what used to be winter and knowing you were going to die. How you'd feel sadness but also a sneaky relief, like you were getting out just in time.

I observed that in timing the birth of her children what she really seemed to be doing was timing their death, and this seemed so premature and absurd—so anathema to the whole framework of parenting—that we both started to laugh.

I hope you're not offended by any of this, though I'd understand if you were. We were mainly kidding, saying things out loud to see how they felt. And to be honest I enjoyed picturing you three times my current age, like a grandfather, as if ancestry were not linear but cyclical and you would one day have a chance to pass your own meandering letter back to me. That way you could have the last laugh ("Despite your carefully conceived plans to the contrary, I am not dead yet, nor do I intend . . .").

Our friend Kurt, who is queer, likes to tease us about these anxieties, calling them our "breeder problems." It's not that he doesn't care about the Problem. He's as terrified and indignant as we are. But these feelings need not take shape around the object of a child, let alone a biological child. (Kurt, for his part, is planning to adopt.) There are many other ways in, many methods and motives for grappling with the crisis.

This letter is only one experiment in the far larger project of imagining people into being. People who exist, or don't, or may never. People we've met, or haven't, or may not. Strangers separated from us by continents and generations.

What I'm describing here is not an outlook so much as a practice. The practice of stretching our attention far beyond its habituated scope, of calling to mind again and again and again the people we've been conditioned to neglect until they actually weigh on us, until they take up residence in our heads. This is maybe the chief ethical and political challenge presented by life in the Anthropocene. To finally match our obvious and increasing interdependence with an appropriate breadth of care.

This is precisely what the Pruitts refuse to do. When a storm hits, they turn our losses into figures, to be tallied or expunged. But to care for anyone is to resist their abstraction, to make them harder to sum up and kill. As cultural historian Saidiya Hartman puts it, "Care is the antidote to violence."

When I started writing you this letter, I wasn't thinking in terms of care. I needed a means to process the Problem, and this was a cultural mode readily available to me as a cisgendered, straight man. Thinking of "the children" was one way I'd been taught to access hope.

The queer theorist Lee Edelman might refer to this instinct as "reproductive futurism," a pattern of thought by which the figure of the "child remains the perpetual horizon of every acknowledged politics, the fantasmatic beneficiary of every political intervention." In his book *No Future*, Edelman argues that queerness should embrace the death drive and reject the cult of the child, which merely serves to reproduce the dominant social order.

He rages against—but also, in a way, *through*—Pope John Paul II's assertion that same-sex unions have "no future and cannot give any to society." "Queers," says Edelman "must respond to the violent force of such constant provocations not only by insisting on our equal right to the social order's prerogatives, not only by avowing our capacity to promote that order's coherence and integrity, but also by saying explicitly what the Law and the Pope and the whole of the Symbolic order for which they stand hear anyway in each and every expression or manifestation of queer sexuality: Fuck the social order and the Child in whose name we're collectively terrorized; fuck Annie; fuck the waif from Les Mis . . . fuck the whole network of Symbolic relations and the future that serves as its prop."

When I first read Edelman, it became clear that the Problem had even further complicated things. Because now the social order was threatening to destroy the future in whose name it was ostensibly arranged. Which meant that in succumbing to my status as a breeder—in having you—I would be simultaneously reproducing it and hastening its self-implosion. I added this reluctantly to my growing list of justifications, both for and against.

PROBABLY HE'S RIGHT. Probably I am too fixated on reproduction, among all the possible vectors of care. It still feels important to try to humanize you, though, if only on the page, if only in defiance of the Pruitts' story, which aims to render your whole generation invisible (along with large swaths of my own). This erasure is not incidental. Its assumption is embedded deep within the financial markets through which our university invested its endowment. In the modeling that informed these markets, it was standard practice to apply a "discount rate," which, to a greater or lesser degree, cheapened the value of future benefits relative to that of present costs. Through this translation, a modest boon to my generation could be made to outweigh enormous harm to yours. The rationale for this was that your generation (and all future generations) would be better able to deal with the costs because you'd be richer and cleverer than us—two premises that rang particularly ironic when used, as they often were, to justify continued investment in the coal industry. But the irony gets even twistier than this, more circular. For instance, it was your dehumanization that made possible the investments that funded my college education. Without those four years of coal-backed writing courses—without the subsequent chain of retreats, recommendations, and fellowships made possible by my degree—it's unclear if I'd be writing to you at all.

I wish I could tell you that this kind of evil was a static thing: that it belonged only to certain people, laced their blood but not mine. But I don't think this is how it works. This evil belongs to no one. It moves and spreads, sweeps through us like a contagion. It can jump from person to person, city to city. It can well up through the gas pump and make your own hands smell like flame.

I can tell when it has infected me, passing from their bodies into mine. I can watch it coming out of their mouths with every televised denial, pouring invisibly onto their podia and across the floor, seeping into their audience and spreading virally outward. It comes up in me as an anger, so consuming that it feels like pain, a single, white-hot point. In its thrall, I compose pointless diatribes in my head, delivering them scathingly to an imagined lineup of Pruitts. I confront them with a rage so piercing and distilled that it burns straight through their lapel pins and their suit jackets, their whole hackneyed armor of justification.

They are the ones who have drowned thousands to make millions, I shout in these fantasies. They are the ones who buried their misgivings and doubled down. They are the ones who silenced the experts and put trust in their own ignorance. They are the ones who invoked the name of their God in order to destroy his putative creation. They are the ones who eroded the humanity of the vulnerable, translated them into externalities so that they could be killed in droughts and fires and storms without jeopardizing the size of returns. They are the ones who would rather upend civilization than upset the narratives that have allowed them to feel—for centuries, and always in error—invulnerable, inviolable, in control. Who have committed perhaps the greatest crime in human history, and with such petty tools: greed, hubris, an infant's moral solipsism.

I recite these polemics in my head, where they burn like a flame, devouring the oxygen that allowed anything else in me to breathe. Let it burn them, too, is how I feel. If the anger cannot be contained, then let it scorch us all beyond recognition, leave our carbonized bodies at the edge of a footpath in the park.

Maybe now you can see how this works, maybe you've already experienced it yourself. How if you're not careful the evil can infect you like a virus. How you can begin to show symptoms of your own. In my head even now, I am conjuring Pruitts just to gut them. The logic of dehumanization is sticky this way, transitive. To push against it is to risk contamination.

But aren't I justified in wanting some resolution, some catharsis within the crisis? If everything's going to hell, then can't we at least make some final judgments before closing the karmic ledger?

More than revenge, more than apology, what I'm realizing I want is an ending. An ending allows you to name things, to recount them. An ending can tie up the loose threads, draw a ring around the story.

Sometimes I suspect that this is what the Pruitts want, too, that this is why they do everything in their power to prevent a solution to the Problem. After all, our end-of-the-world imaginaries contain plenty for them to relish: the government shrivels up, society splinters into ethno-religious tribes, men take up arms to defend them and theirs. A Nietzschean logic of survival and domination becomes not only justified but necessary. Much like in the Rapture, with which I suspect it is often conflated, the imagined ending allows for a conclusive separation of hero from villain, strength from weakness, bravery from cowardice. It's the erotics of apocalypse; it's *Mad Max* and *The Road* and *The Day After Tomorrow*; it's Alex Jones selling Super Male Vitality supplements to his listeners, salivating for a world where only the ripped will survive. Of course this too is a death drive, nearer to Edelman's than they'd like to admit.

But here is the point: We will never get an ending, and no grievance can justify our desire for it. There will be no credit roll, or curtain call, or finalizing conclusion to this slow-motion emergency. The nature of the Problem is to *just continue*, chaotic and inconclusive, sloughing off all the stories we want to tell ourselves about vengeance or courage or salvation. Maybe this is why we hate it so much, why one side reacts in fear and the other in its mirror, denial. We hate it because it has no allegiance to narrative. We hate it because, as Ghosh might say, we do not know how to write it down.

———

IN THE FINAL WEEK OF Earth: Evolution of a Habitable Planet, the entire class took a field trip to examine the layers in some local bedrock. The professor had hired an old school bus, and Emily and I hunched ourselves into one of the brown vinyl bench seats, holding our backpacks in our laps. Thirty minutes away from campus, the bus pulled abruptly to the side of the highway and let us off on a scrubby median at the base of a large roadcut, its shorn face rising eighty feet above the traffic.

We spread out dutifully along the foot of the cliff, clambering half bent up the scree and craning our necks toward the top. Shouting above the traffic noise, the professor pointed out what fossils there were to be found, tiny cones and radials imprinted here and there in the dirt, bringing to mind an irregularly patterned wallpaper. Half visible in the face, we could also make out various geologic strata, discernible only by their subtle distinctions in texture and coloring. The professor kept referring to these layers as "chapters," as if in a story, though they looked duller than this to me, somehow procedural, like folios in an elaborate filing cabinet—like time itself had been compressed into a system of reference, squeezed into the smallest possible space.

In each layer we could see—or at least were understood to be seeing—the traces of a different world. A shallow sea filled with strange combs and scuttling worms, shells plated like armor; then a swamp, redolent of decay, of club moss and fern; then a vast plain scraped naked by ice—each of them separated by millions of years and only inches of dirt, their unique moods and logics reduced to the barely perceptible difference between off-white and grayish-tan, between chalk and clay.

And suddenly, I don't know how else to describe it, it felt like the whole scene was folding, that Emily and I and the rest of the class—our clipboards and packed lunches, the network of highways that had gotten us there, the sandstone hills through which they'd been carved—that all of it was in the process of collapsing into a line in the rock, like we were already in the throes of that endless palimpsesting crush that would eventually file us away into the earth, stripped of all context and detail.

Above us the cliff looked like a catacomb, its face inlaid with the remains of the dead. Behind us traffic whipped through the roadcut: fossils shot through with cars shot through with fossils, time slicing across itself in fast-forward. The bodies of the vehicles shone turquoise and silver in the sun, passing one behind the next, displacing the air into a breeze. And as the morning grew hotter, Emily and I retreated to the shade of the school bus, which had returned to ferry us home. We sat there until it was time to go, until the driver turned over the engine. Then the tail-pipe shuddered to life, issuing nothing we could see.

Second Movement

It is the beginning of 2017. In three weeks, Trump will try to add a zero to the size of his inauguration crowd. In ten months, he'll try to subtract one from the death toll of Hurricane Maria.

I'm a year into my job with NY Renews, and so far both of the bills we've been trying to pass are stalled. Already this was worrisome, but the impending inauguration of a president who has pledged to end all federal efforts to combat the Problem has made their passage seem almost desperately urgent. An email chain has taken off among the young New York organizers who work mainly on the Problem, a list of only a couple dozen people. There is a growing sense that in order to generate coverage and press the issue, we will need to take more direct action.

I am always surprised by how few of us there are; how, even in one of the biggest cities on the planet, all it takes is one large meeting room to accommodate a good portion of the millennials who make a living trying to forestall the apocalypse. And we all know one another—we go to the same parties, trade the same rueful memes and modestly helpful coping mechanisms. Most of us are embedded with one organization or another and spend our days building funder-approved initiatives to fight the Problem, focusing on development and deliverables. Sometimes we feel cynical about this, though for the most part we believe in the importance of our jobs and the nonprofits we work for. But at night we gather to vent and plot outside their strictures, letting our excitement and despair leak out safely among peers.

There is a sense—qualified, self-conscious—that we are part of a historically important movement, and we wear this feeling awkwardly, heft it around as both a prize and a burden. It comes out when we're with one another, exposed in little flashes of ambition and overwhelm. We deliberate over speaking roles to make sure nobody monopolizes the megaphone. We text blue heart emojis to our friends when they're in crisis over the latest scientific projections. We trade books and insights about past movements, mining for inspiration in the words of Bayard Rustin, or Ella Baker, or Myles Horton. We go to happy hours and gossip about who among us might burn out, or go to jail, or run for office.

This is the crowd that comes together after the initial email chain, gathering in the room where I usually take my conference calls. The meeting is held at night—after we've all managed to slip away from our nonprofit day jobs—and out the windows you can see the dark shine of New York Harbor. The Statue of Liberty looks small and redundant amid the lights of passing ferries. Across the water, there are rows of massive gantries casting a violent, halogen glow.

The feeling of these meetings is always the same: agenda in marker on the whiteboard, running notes on a shared Google doc, a round of introductions for the benefit of the few people who don't already know one another. Someone has brought a bag of clementines to the meeting, and little piles of rind accumulate on the table as we talk. We agree to do a sit-in together at the

New York State capitol, occupying for as long as possible the hallway to the governor's office.

The governor has recently been talking a lot about his plan to provide free college for every student in New York, and we decide to pin our messaging here: it is commendable but ultimately pointless to educate all New Yorkers for a brighter future if the government doesn't simultaneously take the steps necessary to protect that future from the Problem. There is also a point to be made about the airports—La Guardia and JFK—which the governor has lately been fixated on refurbishing. What is the point, we will say, if the governor is unwilling to take action aggressive enough to prevent a rise in sea levels that would inundate them both? This basic rubric can be repeated for almost anything: Foregrounded against the trend lines of the Problem, so much of what the state does now can be made to look shortsighted, hypocritical, absurd. There is always an asterisk hanging over their best-laid plans, and it often seems like our job to drag it into the light of public discussion, even as the governor—responding to a mixture of political expediency and what seems to us like magical thinking—prefers to keep it mostly out of sight.

When we wield our talking points, we do so from out of our age. "As young people" we write at the tops of our press releases and the beginnings of our speeches. We rarely feel young when we say this. Our youth just feels like a commodity, something incidental we have to leverage. In the media, we wring it of all possible symbolic import, all moral authority, until the only things left are the numbers of our birth years—1988, 1991, 1997—years that on the timeline of the Problem seem impossibly long ago: back when atmospheric carbon levels were just passing a quaint 350 parts per million, what was once considered the "safe" limit. Even geologically, the era of our birth is distant and irretrievable, and although not even thirty years have passed, biographical chronology seems to work differently for our generation than our parents', with time more prone to frightening warps and accelerations. Whenever I use the phrase "as a young person," I feel old as hell, weary with impossible responsibilities, already a kind of relic.

Still, we are in our twenties, and when it's just us in a meeting, things feel different, looser. After we've put the finishing touches on our plan, there's a rush of inspiration in the room that we may not have let ourselves indulge in the presence of our mostly middle-aged funders and directors. People hug and clasp hands, lingering in the office to look out the windows. For a few moments we allow ourselves to feel the almost embarrassing gravity of it all, unbuffered by mission statements or organizational objectives, a struggle so dramatic and high stakes that it is hard to believe we are part of it.

The weeks following pass as if in montage: recruitment emails are sent out, posters are designed and printed, press releases are drafted, hashtags are brainstormed. Once we have a solid cohort committed, we bring in a trainer—a woman who helped found ACT UP back in the '80s—to conduct a workshop on nonviolent direct action, Skyping in the recruits from Buffalo, Albany, and Long Island. On the day of the training she wears horn-rim glasses, a black turtleneck, and black jeans. Nonviolence is not pacifism, she tells us. Nonviolence is like jujitsu, where you use your opponent's momentum against him. It lures the violence out of the system, forces it to show its face in public. The point you're trying to dramatize here, she says, is that the administration would rather have twenty young people placed in handcuffs than support legislation to protect their futures from the Problem.

After she is done with her explanation, we practice the choreography, linking hands in a circle and sitting down in unison at a signal from one of the people on Skype, whose laptop we've repositioned so they can watch. We do this three times, getting up and sitting down again on the office carpet.

After the training is finished and people begin filing out, the woman from ACT UP puts her boots up on the table and leans back. "I don't know how you guys are doing this," she says to me. "The Problem is just so . . ." She flips her hand in the air, grasping for a word. "Heavy," she settles on, putting her boots back on the ground. "I always forget about it then remember again and I'm like, should I be worried about my kids?" She looks at me seriously for a second, as if she thinks I might give her an answer, then shakes her head and stands to leave.

In the days leading up to the action, we cut out twenty cardboard life preservers using a stencil and paint the words Save Our Future on them in white. Spreadsheets are created for carpools, and other spreadsheets for day-of roles (police liaison, live tweeter, snack purchaser, videographer). We spend hours on the phone with an Albany lawyer who's offered us pro bono legal advice. Several of us take a trip to go scope out the capitol, creating a map indicating the location of the elevators that will lead directly into the executive hallway. The closer we get to the action, the more emails we send, until our phones buzz incessantly like little avatars of our mounting nervousness.

The morning of the action, those of us coming from New York City meet at a van rental lot in Midtown. It is still early, and the guy behind the desk has to open a metal grate in order to let us into the office. Where are you going, he asks us, looking down at some rental forms and checking boxes. To the capitol, we say, for a field trip.

It takes us two hours to drive to Albany, past winter hills and small, sleeping towns. We spend most of the time calling the reporters we haven't already tipped off, letting them know what we are about to do. At the capitol, the two dozen of us meet up in the basement, squeezing as inconspicuously as possible into the appropriate elevators. When we finally step into the governor's wing, the hallway is silent and empty. Both walls are hung with paintings from the Hudson River School, bucolic scenes of pre-industrial New York set into gilt frames. The lone security guard gives us a quizzical look; static-chopped voices murmur from the walkie-talkie on his belt. In the moments before his confusion provokes a response, we walk past him as a pack and sit down in front of the entrance to the hallway.

As soon as we sit it is like an invisible button has been pressed on the marble floor, and a series of interlocking systems spring to life around the building. More security guards show up, rushing down the hall, then the first press photographers, then the capitol police, perfectly calm in their gray slacks and knit purple ties. The corridor is shut down to the public, though curious staffers crane their necks past the circle of guards, trying to catch a glimpse of the commotion.

In the middle of the growing scrum, we unfurl our banner. It tells the governor exactly what bill to pass. It doesn't rhyme or mince words. The press will, we know, look for any opportunity to paint us naive—"demonstrators" they might call us, or even worse "young enviros"—so we never risk being cute. The emphasis has to be on discipline: point the banner at the camera, chant without flinching, stick to the talking points. The governor's not even in his office today, but our hope was never to speak with him directly. We know that if we get through to him at all, it will only be secondhand, through reporters and advisers, the whole bulwark of surrogates meant to shield him from our resolve.

One by one we get up to speak, talking honestly about our fear, demanding that the governor take action commensurate to the scale of the Problem. The shutters on the cameras click. The police shift from foot to foot, hands clasped by their waists. One or two of us tear up and the tears are real, embarrassing though we will them not to be. But still it feels like we are staging a play, something we've rehearsed at length and are now acting out for an audience of reporters and security personnel.

In a basic sense this is true. We've produced the sit-in specifically for the benefit of the press, hopeful that the syndicated image of dozens of young people pleading for their future outside the door of the governor's office will by necessity provoke some response from him, thrusting the debate into the public eye and winning us some leverage.

But it is also true on a deeper level. In organizing daily against the Problem, we've become so adept at compartmentalization that these actions are often our only chance to really grieve. The sit-in creates a context for it, a moment of heightened drama into which we can finally pour our anger and our disbelief and— scariest of all, because we guard it so closely—our flickering but still unextinguished sense of hope.

For this, we are our own audience. We perform our feelings and then they become real enough to wash over us. Around our circle, the machinery of the capitol reacts, and we are satisfied to watch it scramble. We have sharpened catharsis to a point, punctured at least for a moment the bubble of numbness that otherwise shrouds everything, all the time.

Inside this bubble, the news anchors can devote more time to the British royal family than to the threat of rising sea levels; the governor can react with greater urgency to a small-time prison break than to the dire warnings being issued by the United Nations; and people in bars can still nod at us politely when we tell them, with equal parts pride and reluctance, what it is we do for work.

We're reminded constantly that while we think and plan and worry about the Problem nearly every day, the majority of people still almost never do. Being inside the bubble can make us feel crazy this way, and lonely—like we are trapped in one of those classic nightmares, a monster closing in and everyone around us gone deaf. Yet to a certain extent we live in the bubble, too, anesthetized by email, focused on whatever the next task is, the next meeting. The drama is by far the exception: most of our work has us sitting in front of our laptops, or scribbling notes on butcher paper, or printing out spreadsheets and attaching them to clipboards. It is often desensitizing, in other words, and this lends the work a surreal aspect, because with every control-shift and reply-all we are supposedly trying to contain a vast and growing Problem in the atmosphere itself, like our keyboards are somehow linked up with the sky.

This mundanity is, to an extent, unavoidable. Despite the sanitized stories we've been fed on Martin Luther King Jr. Day— the way the man himself has been whitewashed with an especially sapless brand of heroism—we know enough by now to realize that social movements do not spring up out of nowhere, that it took dozens of organizers months of careful planning to pull off the Freedom Rides and the March on Washington. Still, the enormous experiential gap between the texture of our daily work and the violence of the Problem it seeks to address can feel numbing, like we are wading through a dream. Direct action bridges this gap, allowing us for a moment to feel the emotional immediacy the Problem deserves, to see it reflected in security cordons and the lenses of cameras.

After about half an hour the police begin arresting us. The official charge is disorderly conduct. One at a time, they motion for us to stand and strap plastic handcuffs around our wrists. Then they take us by the arm and walk us down to a converted conference space in the basement of the capitol, where we sit in folding chairs waiting to be processed. The whole procedure is polite and pro forma. On the way down, each of us has to have our picture taken with our arresting officer next to a bank of metal detectors. "This feels just like prom," someone says, and everyone in the room laughs.

Eventually we are released, with a summons for a court date the following month. The action has worked more or less as we intended: we've garnered enough press to get a rise out of the governor and will soon parlay this into a meeting with his chief of staff. But by the time I sit down in the van for the ride home, the thrill has largely worn off. I feel tired, hunched in my seat by the window. Unfolding my laptop, I squint into its clamshell glow. For a moment, I am surprised by my empty inbox and then I remember I have no Wi-Fi. I take out my phone and turn on the mobile hotspot. The screen seems to pause a beat, and then the messages arrive, a block of bolded subjects running down past the bottom of the screen, typical for a day spent off of email. There are messages from reporters, and volunteers, and directors of this and that, some of them sent just to me, some to lists of thousands. I work through them mechanically, dispatching each with a sentence, maybe two, unclear on what I am saying until my fingers are already producing the words as if in a fugue state, capitalization falling away, syntax blurring, the replies appearing on-screen like I haven't written them at all, like they are being proffered to me by some barely cogent third party. After a while I close the laptop, unaware of what I've written, unable to make myself care. We are already halfway back to the city.

I spend the rest of the drive with my cheek pressed to the leatherette of the sill, watching the pavement replenish itself frantically in the headlights. It never stops being strange, that this is what I do now, that strategic breaches of the law are a part of my work, that I carpool to and from them in a van. As a child I'd had other plans. At one point I'd dreamed earnestly and absurdly of becoming a professional soccer player, still young enough to believe that careers arose naturally out one's favorite grade-school activity. Later, when my visions for the future sobered, I pictured myself as a landscape architect, raking beds of gravel into gently curving patterns, charting out pleasing constellations of trees.

That was back when the world was a hidden premise, a smooth, hard plane. The future seemed to rise from it like a ramp, pointing upward at a manageable grade. All we had to do was walk up it in the direction of our choosing and we'd come out better off, higher up, with a more expansive view over the terrain.

But in the years it's taken us to age into the job market, something huge has shifted beneath our feet, a kind of temporal tectonics, a jolt of acceleration. Suddenly the world is changing faster than we are, its cycles spinning out of control. Geologic

time—for eons a dull murmur behind the backdrop of history—
has come crashing into the inviolate space we'd reserved for
those myopic little snippets of eternity that we referred to as our
lifetimes.

In the course of a decade, then, it's become vaguely absurd
to build our future in the world like we would a house on some
land, just erect it and trust it to stand. Increasingly, the only viable future seems to be in shoring up the future itself. And so
the world transforms from a premise into a question, and we
work desperately to answer it in our favor.

The strangeness, though, comes from still being able to remember that prior world, the one in which nothing was going
to change, in which it was still possible to treat the future as a
linear projection of the present. The Problem makes childhood
seem very distant this way, a tiny scene at the end of a backward
telescope, made of a different substance entirely than the halls
of the capitol, the plastic of the cuffs. Then again, maybe this isn't
so strange: just an update on that old loss of innocence, the latest expulsion from Eden. Just growing up, in other words.

Loss

I t was in the early 2010s that the storms began to feel like more than just weather; they'd become semi-routinized now, another way to mark out the time. When a storm made landfall, footage would flash for days across every screen in the country, forging from our millions of daily lives a brief slice of historical simultaneity—a reliable "where were you when"—so that eventually, without any conscious involvement on my part, the disasters themselves began to serve as increasingly prominent inflection points in my biographical memory.

The more catastrophes I witnessed, the more a pattern began to reveal itself. It would almost always start with my phone. A news alert would pop up, informing me that a wildfire was gaining ground, or a storm was threatening the coastline. I would monitor its progress, tabulating who I knew in the area, praying for a mollifying rainstorm or a sudden redirection in the wind. Then, assuming these prayers went unanswered, the internet would erupt in a first wave of videos, shaky and coarse-grained, just a few seconds long: a smoldering patio, rain lashing a flooded driveway. After watching a few of them, I would put away my work on our bill package and switch into "rapid response" mode, connecting with other members of NY Renews to work out how we were going to respond. On a dime, we would all try to change tempo: offense to defense, future to present, urgency to emergency.

I remember watching Hurricane Maria hit Puerto Rico on the desktop computer at my office, which at the time was located at the far southern tip of Manhattan, an area that had been submerged during Sandy five years earlier. By then, the coverage of the storms had been ritualized, a thing the news networks knew how to do, stretching it over a forty-eight-hour cycle of worry, panic, loss, and catharsis. Many of the pictures coming through were familiar to me from previous storms: people on rooftops, huddled in lifeboats, wading across intersections strung low with dead stoplights.

Maybe, by the time you read this, Hurricane Maria will have become a footnote—its importance diluted by the accelerating chain of subsequent storms that as of now have yet to occur. But there is one image that's stuck with me enough to resurrect for you. It showed a row of pastel-colored homes on a rocky beach, their facades torn away so you could see into all the rooms, as if peering into the innards of a dollhouse. The edges of the houses where the front walls had collapsed were a shrapnel mess of broken glass and splintered wood. But inside, the rooms themselves looked strangely untouched, like curated memorials to the lives they'd only days before contained.

By the time I minimized this picture and opened my inbox, I'd already received a flurry of emails about organizing a vigil for the next day. No one was articulating a strategic rationale: maybe the New York TV stations would pick us up as a local coverage hook, but this felt like a moment when mourning could be an end in itself, when the communal need for some form of response was enough to justify the hours of work it would take to pull one together. We spent the rest of the night working furiously, making calls, blasting listservs, formatting a Facebook event, drafting a press release, buying candles, printing posters, lining up speakers, identifying organizations on the ground to whom we could direct donations.

Someone had found the office address of ExxonMobil—the oil company that had recently been indicted for deliberately hiding evidence of the Problem for decades, including research predicting an increase in the intensity and severity of hurricanes— and it was there that we chose to gather the next day, reminding ourselves and anyone watching that the tragedy had a source, an etiology beyond coincidence. The building was imposing and nondescript, a Midtown skyscraper with a cavernous lobby and a fountain in the plaza out front.

About eighty people gathered near this fountain in weather that threatened rain but never delivered. Some held candles and others held signs. "Solidarity with the Victims of Maria" read one. "Exxon Knew" read another. The mood was somber, but nobody cried. Pedestrians walked by in a hurry. The fountain threw off a fine mist that dampened the posters and guttered the candles. I spent most of the time broadcasting video of the small gathering on social media, hoping in vain that it might take off, that our tiny memorial would help inspire a larger outpouring on the internet.

After it was over, a woman who'd been standing in the back came up to me and confided that this was technically no longer the Exxon building, that they'd relocated their office years ago. That night, because of the train delays, it took me two hours to get home, rocking with the subway as it stuttered through the

city, both pockets full of extinguished candles, which I'd collected after the vigil and forgotten to throw away.

I'm sharing this with you because I want to convey what it was
like to feel that we were losing. To look up at the glass face of an
arbitrary skyscraper and know that the storms were getting
worse and that thousands more were going to die. To know that
many people were working very hard to prevent this—legislators,
engineers, artists, activists—and that these efforts were, at least
so far, nowhere near adequate. And to keep trying anyway, to
tread the liminal ground between denial and resignation, not
always buoyed by hope so much as the terror of what giving up
would force us to admit to ourselves.

Whenever I felt this way, I had to fight the sensation that we
were brandishing our posters into a hurricane. That the corporations we were up against—invisible in their office towers, mobile across borders and markets—were simply another form of
weather, as natural and untouchable as a storm. For me, this was
a new lesson in losing: that it wasn't enough to keep contesting
your opponent, that you also had to guard against your own desire to hand them the mantle of inevitability, to say, simply, that
there was nothing to be done.

Later it came out that the storm had killed the grandmother of
a friend, an organizer from Buffalo who worked closely with the
coalition. In the weeks following we held a rally at the capitol—
five thousand people came this time—commemorating the dead
and demanding that the governor pass our bills. Almost a year
had gone by since the sit-in outside his office. Coverage of the
arrests had brought the administration to the table, but they still
hadn't made any real commitments. Marching down that same

corridor, we had to fight the sensation that we'd been walking in place—that all that time had been wasted.

Our demands hadn't changed in the intervening months: it was past time to decarbonize the economy; to tax the big polluters and use the revenue to protect storm-vulnerable communities. My friend spoke through anger and tears to tell the crowd about his grandmother, who had died in her home in San Juan. I remember that I put my hand on his shoulder as he spoke, but I couldn't bring myself to look directly at him. It felt unbearably private, this struggle to translate personal grief into collective action, speaking sadness through a megaphone and hearing it echo down the hallways of the capitol. All I could think was that he shouldn't have to be doing this, that it was too much to ask, that I'd escaped this responsibility only because my own grandmother had been spared by her storm.

The nature of the Problem made it so that we were grappling constantly with this question of when and whether to turn tragedy into strategy. With each storm came the challenge of organizing a response that balanced mourning and mobilization, something that carried the weight of a memorial but also worked to avert its repetition. Sometimes we got it wrong. Sometimes we felt like the worst kind of ambulance chasers. And there was always the issue of where to put our grief, where we could lay it down so that it wouldn't get trampled in the commotion of our response.

————

WHEN I FOUND no other place for it, I funneled my grief into work. I worked harder, clocking more hours, diverting more thought. My grief tethered me to my phone. It responded to a rash of emails in the morning and another one before bed. It logged out of one conference call and immediately into the next. It signed everything "Best," which it knew was disingenuous.

I was far from alone in this. Everyone I knew who was working on the Problem threw themselves at it with sub-healthy abandon. It had an incandescent gravity, like a lightbulb for moths, something painful we kept slamming ourselves into. Many organizers spoke about the importance of self-care, of taking days off, of bringing calming herbal teas to work. But even this struck me as an instrument of the broader logic: *do what you need to do to ensure you can keep working hard in perpetuity.* The goal was to push as much as you reasonably thought you could without getting so tired that you gave up completely. Many of us operated right on the edge of this line, constantly trying to fend off burnout—a phrase whose irony wasn't lost on us.

It would sometimes get bad enough that my first reaction to news of a new disaster was not worry but exhaustion. Exhaustion that, after a long day of campaigning to pass legislation mitigating the Problem, I now had to steel myself for many more hours trying to respond to an event caused by the widespread failure, as of yet, to pass precisely this kind of legislation.

This reaction was usually accompanied by a commensurate guilt: that I was losing sight of the lives that were at stake, that I was letting my fatigue get in the way of my empathy and my efficacy. That I wasn't working hard enough, basically. This would prompt predictable thoughts of all the millions of people who on a daily basis had it harder than I ever would: the Bangladeshi farmers trying to coax crops from increasingly saline soil; the Sahelian women who had to trek farther and farther for a single jerry can of semi-potable water; the Marshallese who were already coping with chronic flood tides, saltwater washing through kitchens and breaching shallow wells.

With this thought, I could always muster a second wind, and a fifth. It was a tiny sacrifice, after all, to have to sit in my comfortable desk chair a few extra hours, to return home late to a cozy apartment near an as-yet-unflooded stretch of coast. Truly nothing compared to the real-life tragedies I knew were unfolding constantly in the peripherals of my worldview. In this way I was able to transmute sadness into work and back again—a loop that's kept me going for years, one I'm not entirely sure how to escape.

Sometimes when I'm freaking about the Problem, a friend will tell me—maddeningly, prophetically—that I need to "chill." On these occasions, my guilt cuts the other way. They're right, I think: "Chill" may be the only defensible posture now. What we need, according to many analyses, is to work *less*—take fewer trips, make fewer transactions, stay home. Slow the metabolism of the global economy way down.

It's honestly a pleasure to envision this new world of loose deadlines and long afternoons. There'd be more boredom, more silence, more happenstance. You would wake up one day and there'd be nothing on the calendar at all. Maybe you'd put on a pot of tea, maybe you'd take a walk to meet a friend. On the way you'd see lots of other people, out for the same reason you were, taking unhurried strolls to vague destinations, stopping now and then to tie a shoelace or chat with a stranger.

In this other world, laziness would be the exalted ethic. Fuck tilling the earth, and building the brand, and making your mark. Fuck spending your time, like it was a pile of unused cash. Fuck the Protestant ethic, that congenital masochism, which I've inherited even from my Jewish side, who'd had it handed to them at Ellis Island with their meager effects and their bastardized surnames.

Even generations later, a part of me is still falling prey to the myth of salvation through exertion. I know the belief is misplaced, but I haven't managed to shed it.

It's why the only response I seem capable of in the face of a Problem born of too much work—of a literally overheated economy—is to make myself work more. This is the strand of that other guilt, its persistent voice. How can I expect the economy to slow down when I refuse to do the same?

There was of course a moment, at the beginning of the COVID pandemic, when the economy really did seem to slow down. Skies cleared up and offices cleared out. Plane travel plummeted and global emissions dipped modestly. As people around the world took shelter, watching anxiously for news of the virus's spread, animals not seen in decades began venturing back into cities and suburbs, emerging as if from their own compulsory quarantines.

Soon, though, something else seemed to take hold. Essential workers went back to work, accepting more risk for the same pay. Those lucky enough to keep their office jobs were forced onto a treadmill of eye-glazing Zoom calls, while those who weren't were left largely to fend for themselves. Unwilling to spend the money to support working people during lockdown, government leaders rushed to "reopen" their economies, paying for their haste in the currency of human life. In the United States, where this failure was most parodic, hundreds of thousands of people died. The massacre was the worst in poor and nonwhite communities, where social services, medical care, and wealth accumulation had been winnowed down for decades. Meanwhile, evictions continued apace. So did Black Friday, standardized testing, and mandatory minimum sentencing. The system lurched forward like a shark, unable to slow down for fear of dying.

I think I still believe in the promise of the lazy economy, but the pandemic underscored how much would need to change to make it real. Old convictions would have to be not just abandoned but inverted. Distribution would have to supplant accumulation. Abstention would need to replace industry. Degrowth would need to be the ultimate goal, deceleration the means by which we got there. Maybe then, once we'd finally learned to relax, a collective chill could cut the rising heat.

I remember a point, a year into my time with NY Renews, when I started taking my conference calls outside. This was a concession to my friends' advice, a half-hearted attempt to access the chill I didn't have. I'd plug in my earbuds and go for a stroll through the neighborhood, pretending to myself that I wasn't working, even as I fielded questions on banner design or tax policy.

My usual circuit led me past the site of the World Trade Center and its memorial to those killed on September 11, 2001. The memorial consists of two large holes where the twin skyscrapers used to stand. Discrete jets have been installed along their edges, so that their walls form a perpetual waterfall, sleek and uniform as granite. Regardless of where you stand around the perimeter, it is impossible to see the bottoms of the holes, creating the illusion of water falling forever into an infinite space. Circling the falls is a ring of bronze parapets engraved with the names of the dead—2,997 people in all, almost exactly the same death toll as Maria.

The memorial is located in a hundred-year flood zone, which means that, in any given year, it has a one-in-one-hundred chance of drowning. On a rainy day it wasn't hard for me to imagine this, the next hurricane spilling New York Harbor over the lip of Lower Manhattan. The water would surge across the plaza and pour down into the holes, turning the waterfalls into roaring cataracts. And perhaps the symbolic depth of that previous tragedy would swallow all the water like a drain and save the rest of the city, but probably the holes did in fact have bottoms, meaning the water would eventually overwhelm them and then continue rising.

This was a scary thought, that the Problem could produce one tragedy while erasing the memory of another, and that as it got worse the storms might clog against themselves, leaving progressively less time for mourning, eventually becoming so frequent that the memorials themselves would have to be built and fortified with the next disaster in mind. This became another way I justified my constant work: that I was protecting grief itself, preserving the time and space necessary for its ongoing expression, though between conference calls and email, I could rarely slow down enough to indulge in it myself.

———

DURING THE PRODUCTION OF OIL and natural gas, any leftover fuel that can't be pumped back into the processing system is burned off in a flare stack. I used to see them while driving down the New Jersey Turnpike, rows of giant metal candles that never went out. At night they burned a vivid, medieval orange, though during the day their flames were translucent; what you noticed instead was the air around them, which shimmered and melted like wax. This was the only sign that excess fuel was being wasted, vented transparently straight into the sky.

Flaring is a means of dealing with surplus. When you have too much of something, a flare can make its substance invisible, divert its energies elsewhere. During the years I lived there, New York felt like a city of flares. The billionaires flared excess capital into Basquiats and pieds-à-terre. The pundits flared popular resentment into cathartic invective on Twitter. The police flared white anxiety into targeted harassment in Black neighborhoods. The millennials exhausted themselves pedaling in place in boutique fitness classes, flaring off calories that had accumulated at the winning end of the global supply chain. And whenever I found a spare moment, in a wait between subways or a lull in conversation, I had to stop myself from taking out my phone and flaring it automatically—into email, into podcasts, into any input that could consume the minutes, destroy them.

When they became too much, even fear and grief could be flared, their appearance disguised or displaced. In the heartland, farmers held hushed meetings to troubleshoot worsening disruptions to their growing season, even as they chose leaders who decried the Problem as "junk science." In gated communities along the Gulf Coast—red on the electoral maps, blue on the flood maps—insurance coverage eroded with the shoreline and families jacked their bungalows onto ten-foot concrete pylons (entire suburbs built atop unspoken understandings, like something out of Calvino).

Once, at a bowling alley, I saw a televised baseball game

interrupted by a commercial for a certain brand of home generator. All at once, the screens erupted in frantic imagery: trees bending in a gale, power lines whipping like jump ropes, a city blinking into darkness block by block. An older white couple—sensibly dressed, Cialis-ad handsome—cowered on a sectional sofa in a well-appointed living room, the windows behind them jagged with lightning. I could see their furrowed brows and neat haircuts suspended above every lane, tessellating down the row of televisions.

A male voice intoned vaguely about "catastrophic storms" and "record-setting blizzards," though the Problem was skirted, never mentioned by name. Then the scene switched to another white family, arrayed in a configuration meant to convey domestic harmony. The mother was in the kitchen supervising some sort of electric mixer, the son was in the living room playing video games, and the daughter sat at her desk in front of a laptop. Suddenly, the power cut out and all three recoiled from their devices, indignant and confused. They stumbled around the darkened house, cupping flickering candles in their disembodied hands. They looked weirdly spectral, as if haunting the lives they'd occupied only moments before.

Then the music brightened and the generator came to the rescue. There it was, rendered in digital cross section, rotating authoritatively in front of a neon grid. Its components swept in from off-screen and locked together in a complex blueprint, inscrutable and reassuring. Then the mother pressed a command on her phone and the thing whirred to life, illuminating the house once again. The family hugged and returned to their devices; the daughter plugged in her laptop, the mother placed something in the microwave. Outside, the rain lashed the windows, but the house was now shining, impervious. The name of the generator appeared on-screen, along with the website where you could buy one: www.neverfeelpowerless.com.

———

I SHOULD TELL YOU that even after Hurricane Maria—after images of the devastation had been broadcast to every major news channel in the country, after my friend had eulogized his grandmother in the marble halls of the capitol in Albany—the governor continued to oppose our two bills. He had to do something, if only to bury our rally in the press, so he proposed a ban on plastic bags, touting this to reporters as if it was some sort of ecological panacea.

This was often how it worked: the performance of action to abate fatalism, the willingness to do everything except what was necessary. Morale could be maintained even as its validity was undermined, and politically this math was hard to upend.

There was a whole category of leader like the governor, leaders who acknowledged the science but couldn't seem to digest the magnitude. It was the seventh bullet on their policy platform, an added sentence in their stump speech. It was a forceful bit of rhetoric, though never quite center stage, never the hill on which they would plant their flag. They never did *nothing*, of course. They found marginal ways of cutting emissions, policies that were just good to be flaunted in press releases and campaign materials. But seldom was there an acknowledgment that good did not equal adequate, that the Problem had a threshold and a deadline and that we were nowhere close to abiding by either.

You've read the research, we'd say to them, sitting on the fancy couches in their offices. So where is your desperation? Where is your urgency? When you're driving off a cliff, 40 degrees won't cut it—you need to turn the wheel 180.

This is a democracy, is the reply we usually got. All I do is hold a mirror up to the voters. You ask me about my urgency, but where, exactly, is theirs?

NY Renews had experimented with a typical array of tactics to try to move this dial. Beyond the sit-in and subsequent arrests, there were editorial blitzes, celebrity endorsements, commissioned studies, petition drives, television appearances, and high-level meetings with the administration. But even after our rally in the wake of Maria—even as public outrage began to swell to a level that bold leadership might have coaxed into a mandate—the governor refused to budge. He stayed silent on our legislation, occasionally throwing the kind of bone that allowed him to claim—not entirely falsely—that he was addressing the Problem in increments.

After months of this stalling, the coalition made the decision to pivot to the state legislature, formally introducing our bill mandating full decarbonization, despite having no indication that the governor would ultimately support it. At the time, the New York State Senate was subject to a bizarre arrangement in which eight elected Democrats chose to caucus with the Republicans, lending the latter the juridical majority in spite of their numerical minority. Our goal was not to pass the bill—we knew this was hopeless with the Republicans in control—but simply to generate enough sponsorships to foist public pressure back on the governor, who alone had the political power to push through a policy of its magnitude.

Ironically, the only senators in a position to help us collect a sufficiently impressive number of sponsorships were those same pseudo-Democrats whose cynical power-sharing agreement had compelled our bank-shot strategy in the first place. Because they straddled the party line, the pseudo-Democrats had a reputation for being the only conference able to sometimes get both true-Democrats and a handful of Republicans onto a progressive bill, though by the same token, the minority-majority they'd handed the Republicans all but guaranteed that said bill would never actually be brought for a vote, and that any glimmers of bipartisanship would quickly dead-end.

Still, in the interest of momentum, we held our noses and approached the pseudo-Democrats with our legislation. The conference conferred, and after weeks we were notified that they'd given it to perhaps the weakest of their eight members: Senator A, an aging fixture from the outer suburbs of Queens, notorious for trumpeting visionary legislation and then sitting on it, sometimes for years.

Senator A set about praising the bill, referring to the Problem as a matter of "grave importance" and issuing supportive press statements on his office's letterhead. He also went in search of a critic, however peripheral, with whom he could sabotage its progress. He found one in a retired electrical engineer who'd written an ill-informed critique of the bill's drafting, one that misapprehended the legal application of various words. In private meetings with us, Senator A would keep circling back to these critiques, referring to them, too, as a matter of "grave importance," one that would require months of study and deliberation to address. In this way, he could stall the bill almost indefinitely and have this stalling appear to be a product of his *over*concern for the issue.

Meanwhile, relations between the pseudo-Democrats and the Democrats were growing increasingly bitter. The latter accused

the former of flouting the will of the voters to prop up an ille-
gitimate conservative majority. The former accused the latter of
being unwilling to work across the aisle to get things done, holding up our bill as proof of their progressive credentials, even as
Senator A quietly buried it in legislative purgatory. Soon, the
orders were given from the leaders of each conference not to
sponsor any bills coming out of the opposing conference.

Our bill was suddenly off limits to its strongest supporters,
relegated to the conference that saw it mainly as a public relations
tool. It was at this point that we decided to approach Senator B,
a solid progressive from Manhattan, to see about sponsoring a
version of the bill through the Democratic conference. If both
conferences took ownership of effectively identical bills, we reasoned, then it would be hard for them to back away from that
support when and if the two conferences reunified, at which
point the onus would land squarely on the governor.

Senator B was enthusiastic about the idea but, before anything could move forward, Senator A caught wind of the plan.
Increasingly desperate to maintain his ownership of the issue
and declaim his uncynical support for our bill, he scheduled a
hasty appearance with one of the big statewide radio stations,
an appearance which backfired when he was unable to explain
almost anything about the bill's contents, let alone the nature of
the critique he was still allegedly studying. "The issue of climate
control is one of my biggest priorities," he kept telling the host,
as if he were not a senator addressing the Problem but an HVAC
technician fixing an air conditioner.

In the run-up to Senator B's introduction of the doppel-
ganger bill, we tried delicately to play matchmaker. We made
entreaties to Senator A's office, inviting him to speak first at the

press conference, arguing that the urgency of the Problem could form the basis for a renewed cooperation.

Instead, just before the event was scheduled to begin, Senator A issued a memo attacking Senator B and NY Renews. "This is an issue of grave importance," it read, "and they continue to play politics."

Even several years later, this is still maddening to recount—not least because it is so painfully tedious, a bad satire written in fine print. And there are a thousand more little intricacies that I glossed over: delicate meetings and fractious emails, a headline here and there, pointless extra twists in the larger Gordian knot.

This chaos and intractability extended far beyond New York. In jurisdictions around the world, the Problem was testing the seams of our politics, cleaving parties and toppling governments. Attempts to "tackle" it got set up, then watered down, twisted into insoluble tangles, the policy lost in rhetoric and the rhetoric lost in words, words that—in systems of this complexity, with so much information swirling at such an epistemological remove—often came to replace arguments as the main political fulcra. Semantic debates raged about a "carbon tax" versus a "carbon fee" versus "cap and trade." "Clean" energy could mean something other than "carbon neutral," which might be different still from "renewables." Some people wanted a "Green New Deal" while others avoided the term assiduously, clinging to careful messages about "protecting our air and water." Words became anchors in this chaos, talismans of position and identity in a system the whole of which none of us would ever see firsthand. And though there were usually substantive differences between the concepts limned by these terms, most of us simply couldn't understand them all, didn't have time to familiarize ourselves with the mushrooming body of primary research in climatology, geology, ecology, economics, history, sociology, engineering, and political science that ostensibly, beyond any single person's comprehension, illuminated the direction in which we should be headed.

Witnessing the turmoil that the Problem had unleashed in our politics, it became possible to see our predicament as both under- and overdetermined. Under- because each twist in the knot of our failure seemed so random, so tantalizingly contingent. If you looked closely, you could get consumed by all the what-if's: What if this bill hadn't been given to that senator? What if this company had been subpoenaed a decade earlier? What if those first seams of coal hadn't been found beneath an imperialist country in thrall to Cartesian rationalism? The chaotic system implied its alternatives, all the paths it could have taken but didn't.

And yet the sheer weight of human frailty wound through the knot seemed to point at some compounded inevitability, something fundamental about our nature that not only predicted but *necessitated* the Problem. As if the Problem was just a part of who we were, an individual property emergent only at the scale of civilization.

Squeezed this way between fate and its counterfactuals, it was easy to feel at a loss, which for me was the same as feeling loss itself. The leaders of NY Renews tried as best we could to game out every move, but there were always unpredictable players crowding the board: ourselves, our opponents, the planet itself—entities of such breadth and complexity that their interactions were impossible to forecast with any certitude. So even though we thought them through carefully, every strategic choice we made—every sit-in, every rally—was to a certain extent arbitrary. Which meant that when our bills came aground on ego or anger or the imprecisions of language, we really did risk feeling powerless.

I am telling you this because there will be days when you too will feel powerless; when the Problem seems so insoluble and your plans so inadequate that you might be tempted to just accede to the storms, to stay at home and watch them worsen on TV. In those moments, the story I want you to have, the one I want you to be able to turn to, is not the easy one. It is not: we had the answer, and here it is. It is more like: we didn't, either, and kept trying.

In the end, in this case, our trying was vindicated. It took weeks, but eventually Senator A realized that he couldn't believably champion his own bill while opposing Senator B's identical bill, so he withdrew the original and grudgingly co-sponsored with the Democrats. Months later he was voted out of office, along with most of the other pseudo-Democrats, and the Republicans lost their pseudo-majority. With a path to passage suddenly open, NY Renews pushed like mad to get the bill to the top of the agenda. And almost two years after Maria—a full seven after Sandy—the legislation passed into law. New York State would be required to decarbonize its economy over the next three decades, supporting workers and low-income families across the transition.

Once it'd become clear that the momentum was there, the governor rushed to claim the bill as his own, watering down some aspects of our original and saddling it with the ungainly title of "The Climate Leadership and Community Protection Act." He remained noncommittal about the carbon tax that the coalition had proposed as its complement. Still, the victory made headlines.

Vox described the bill as the boldest climate plan in America; *The New York Times* called it one of the most ambitious in the world.

I wish I could ask you how these stories feel to read. Does it help, knowing there were victories? When I tell you we felt triumph, does any of that reach you?

I am worrying that I've made all of this sound too difficult. I wouldn't blame you for feeling discouraged from any work involving the Problem: I know the stakes are enormous, and the odds are lengthening, and the prospect of fighting against them can seem daunting if not Sisyphean. But I also want to tell you that I knew nothing you don't and did nothing you can't. I just took the Problem seriously and then continued to check small things off a long to-do list. This could be satisfying, even joyous.

Though coupled with this joy comes the fear of complacency, that pacifying sense that "it all worked out in the end." The truth is that it almost didn't, that it very well could not have. Senator A might not have been defeated; the Republicans might not have lost power; the governor might not have acquiesced; the bill may still go unenforced. And though I wish beyond anything that I could tell you you'll be fine—that you'll win out if you just keep fighting—I can't bring myself to do that. That promise would ring false, like the ad for the home generator, where the flick of a switch could keep the storm at bay, and a happy ending could be bought on installment. Given the world you've grown up in, you probably understand this already, maybe better than I do: that a well-lit house can't keep the sky from darkening. That resisting the Problem does not mean scrambling for some private vantage above the public floods. Resisting the Problem means acting in the play of politics—with all its triumph and farce—

even when there are no guarantees it won't reveal itself to have been a tragedy.

———

THERE IS A SONG that I remember listening to repeatedly in the years I worked for NY Renews. On the subway to work, at my desk in the office—I played it so often that it's since become a mnemonic gloss, the tune through which I now recall that entire period.

The song is called "Truth" by the jazz saxophonist Kamasi Washington. I was not, typically, a jazz listener. Whenever I heard it played, my primary response was usually mild frustration, a sense that I lacked some quality critical to appreciating its coy little phrases (patience, probably). But something about "Truth" gave me shivers. It seemed to embody a distinct plotline, a formal proposition.

The song begins by layering in its components one by one. First the piano, lounging and sad, then the drums, then the electric guitar gilding the top, and the saxophone jumping over all of it, as phonetic and expressive as a human voice. After a few minutes, the choir joins in, too, a warm dirge-y gospel, a wordless wave. The instruments start easing into one another, getting comfortable. You can feel them find their harmony and revel in it: a sound that's somehow both melancholy and propulsive, like a car speeding through a moonscape. It continues like this for what feels like forever, succumbing to its own trance, a fugue of gorgeous repetition.

Then, six minutes in, something happens. The song overtakes itself. The horns stumble into the drums and the drums punch through the choir. The harmony begins to overheat, shooting off sparks and cracking at the joints. The guitar rears up and abruptly dies; the saxophone lets out a wail that becomes a screech— gasping and sustained, like it's suffering real pain. The song careens further and further out of control, a beautiful, intricate thing consumed by its own momentum. The screeching gets so piercing that for a whole minute the song is almost unlistenable. Whatever room you're in is filled with its imploding dissonance, shards of noise lodging deep in your ears. And then, suddenly, the music collapses and the noise evaporates. The song has burnt itself out.

In those days I saw this basic pattern everywhere. This *was* the Problem, I thought. Or at least, this was the truth of its etiology. Chaos emerging from order. Humanity consumed by its inventions. The rhythmic oscillation between buildup and breakdown, between hope and hopelessness.

With "Truth" playing on loop, I'd wander the city near my office, where the skyscrapers sat level with the harbor. On the corners, expensive coffee shops sold exotic beans from drought-prone countries. People in suits waited at every crosswalk, talking intently to the buds in their ears. Volunteers with clipboards requested small-time donations for worthy causes. Look at what we've built, I would think, looking up at the crest of the skyline, and its magnificence seemed all the louder in prelude to collapse.

You can find despair in this pattern—the randomness, the cacophony—though beneath it, I think, there is also a form of faith. That anything can still happen. That no matter how many times you hear it, truth is always unpredictable. The only certainty being that if we stop playing, the song will end.

If you ever get a chance to listen to it, you'll notice that

"Truth" doesn't end at the breakdown. It appears to, at first, and then out of the silence the piano picks itself back up again. Then the guitar, then the sax, inching back toward cautious harmony, getting louder in the final minutes. And whether it's a new song or another cycle of the same song, there it is. You can't help it. You still want to listen.

The philosopher Timothy Morton describes the Problem as a "hyperobject": an object "so massively distributed in time and space as to transcend localization and elude specification." In this understanding, the Problem is both a mist and a monolith—it is everywhere but it cannot be touched, and so resists definition. When I began writing this letter, I thought that if I could glance the right scenes and themes off its sides, I might begin to illuminate the whole, help us both get a grasp on it.

But after plying you with gas flares and jazz scores, TV ads and conference calls, perhaps you've begun to realize what the problem is. The problem with the hyperobject—the problem with the Problem—is that it has too many sides. Glancing off its endless facets, I worry that all I'm doing is re-creating the general mishmash, an arbitrary sampling of the always-already changing world. It can start to feel like I'm just stress testing metaphors—tapping them because they're hollow, listening for some resonance. (Case in point, another dud.)

———

I REMEMBER THE DAY the Paris Agreement was signed, that fateful accord that was meant to save the world from the hyperobject. I had made the dubious decision to spend the weekend at the vacation home of a rich friend who was celebrating her birthday. I would like to say that I'd accepted the invitation to distract myself from the suspense of what was happening in Paris, but in all honesty I hadn't even remembered the talks were concluding that weekend, so completely did I allow myself, on my few days off campaign work, to bracket the reality of the Problem, as if it were all happening in a dream.

The house was large and sparsely decorated, too big for the ten or so people who'd been invited. We all sat around on the living room carpet getting drunk slowly, our wineglasses balanced on leather ottomans. Out the front windows you could see the Atlantic Ocean getting darker as the sun set and we made our way through another bottle. At some point someone asked me what I thought of the goings-on in Paris (my work meant I was often the one asked these kinds of questions), and I responded evasively, trying to hide the fact that I'd completely forgotten they were happening.

The next morning, I woke up and decided I needed to get back to New York. I felt anxious, like I was neglecting something I was responsible for. Guilty, too, that I'd taken a weekend off, allowed myself to lose track of the Problem. It seemed suddenly unconscionable that I would spend what was perhaps the most consequential day of the decade in a seaside mansion full of acquaintances. When I told my friend that I had to leave early to watch the end of the Paris talks, she looked nonplussed. "Can't you watch them from here?"

"I'm just worried I might ruin the party if the talks go south," I said, realizing as I did that this was actually quite likely.

Another friend of hers—a woman I'd only just met—offered to drive me to the bus stop. Grateful, I went upstairs to pack my bag, eager to leave but trying not to look frantic. By the time I'd said my goodbyes and stepped out onto the gravel driveway, the sun was all the way up and the Atlantic was shining. The weather was absurdly warm for a Sunday in December—almost seventy degrees Fahrenheit, a joke temperature, an irony so overt it could not be entertained. We stood there for a moment sweating into our coats, too disoriented to take them off.

"I'm sorry again for bailing like this," I said. "No, no—I get it," the woman said, as if she wished she were leaving, too. We drove in silence, listening to her phone tell us where to go. At the bus stop she got out and we exchanged a wordless hug, the way fear can conspire to bridge the gap between strangers. Then the bus came and I got on.

On board the Wi-Fi was functional, a minor miracle, and I pulled up a livestream of the talks, which were being held in a plenary hall the size of an airplane hangar. At the front of the room, people in suits were giving speeches from a dais, backed by a wall covered in the logo of the summit, the twenty-first in a series of so far unsuccessful UN meetings to address the Problem. Diplomats with vexillary lapel pins shuffled paper and bent microphones toward their mouths, a whole procedural ballet that seemed only to gesture toward the real discussions happening out of range of the cameras, which kept swooping across the assembled delegates, then zooming in on whoever was speaking. The coverage would sometimes cut away to a panel of commentators, who kept saying, as if scared to jinx something, that as of now things were looking OK, that nothing major had blown up.

I kept looking around the bus in the delusional sense that I might catch someone's eye, share in the tentative excitement, but the few other people on board were sleeping or listening to music. When we finally pulled into Penn Station the feed cut out and I rushed back to the apartment of my then girlfriend, sure that history would be made while I was underground on the subway.

But when I got there, she already had it up on her computer

and the final speeches were still being given. Inside the plenary hall on the screen of her laptop at the edge of her coffee table, pixelated diplomats stood up and looked around, rushing down the aisles to whisper in each other's ears. I had the perverse sense—cuddled next to my girlfriend on her couch, the two of us nervously sharing a beer—that we were watching the final minutes of a kind of alternative Super Bowl, a huge spectacle implicating everyone through their screens, something that could be won or lost in real time.

And then suddenly a short procedural vote was taken and the whole thing was over. The tiny people on the laptop were cheering, banging gavels, pumping their fists in the air. Two women who'd helped lead the negotiations were embracing on the dais, rocking back and forth and laughing. It struck us that we'd never seen televised diplomacy elicit such a release, never seen a room full of technocrats accommodate anything as raw as the joy erupting on-screen, and we couldn't help it, we started tearing up, too, cheering like we were in the room.

Nearly every country had adopted the accord, even the most recalcitrant: Russia, Saudi Arabia, the United States. That the agreement fell far short of what was needed we would find out later, in the days following, reading through the online summaries and interactive graphs visualizing the new emissions reduction commitments. It was a sign of just how far behind we were on the Problem that an agreement whose perfect fulfillment would still all but guarantee the inundation of half a dozen small island nations was nonetheless greeted, in most circles, as a historic if qualified success. Over time this would produce a sustained dissonance: on the one hand wanting so badly to witness and celebrate progress, however imperfect. On the other knowing that each imperfection meant consigning thousands of people to suffering; that every concession was a death sentence for somebody. Hope always cast a shadow this way, masking whoever wasn't illuminated in its vatic glow. I've tried to remember this, tried to leave space for the people being left behind with each ostensible step forward, but that night I failed, got swept up entirely in the tide of optimism streaming from Paris to my ex's living room.

Among the many celebratory speeches that followed the official adoption, one phrase stood out to us. A senior diplomat rose to the microphone and in a booming voice said, "Long live the planet," and it was so melodramatic, so cinematic and surreal, that in a delirium of catharsis we found two white T-shirts and scrawled the words on them in Sharpie. Later her roommate came home and found us both stretched out on the couch in our bizarre shirts. "Did you make those?" he asked. Like me, he hadn't really remembered the talks were happening.

For months afterward we still wore the shirts when we were sleeping or exercising. Sometimes we'd meet for a run around the reservoir in Central Park and realize we'd both worn them, the jagged letters faded a little from the wash. We looked very weird jogging in them, like members of some obscure and unhinged cult. But we didn't take them off. It was like we wanted to memorialize the event on our bodies, remind everyone we ran past of its importance. Looking back, this was probably a subconscious bid for some kind of assurance. As if we could enforce the accord ourselves, enact it purely through insistence.

———

IT WAS ALMOST TWO YEARS before the United States withdrew from the treaty. By then Trump had been elected with a minority of votes and was systematically destroying all attempts to contain the hyperobject, like a hijacker who'd wrested the controls from the pilot and was flying the plane straight into the ground—except with many, many more lives at stake. As with everything, he did his utmost to turn the withdrawal into a ratings spectacular, building suspense for weeks by hinting at various courses of action, before finally announcing, in a televised address suggestive of a season finale on the darkest reality TV show in the world, that he would be withdrawing completely.

When the announcement was made, I was up in Buffalo meeting with several coalition organizers, among them the friend whose grandmother had been killed during Hurricane Maria. The announcement was playing on mute in one corner of their office, but no one could bring themselves to watch. On-screen Trump looked smug and bored. Pruitt stood behind him like a mannequin, cinched up into his tie. As soon as the speech began, my friend walked out of the room. I caught the first few minutes, then turned back to my pile of email, only reading the transcript once it was over. I could feel the farce congealing over the tragedy, numbing it down. I decided to take a walk around the block to see if I could cry.

I came upon the lot of a small church—deserted on a Thursday—and sat down on one of the concrete strips bordering a parking spot. My sadness felt more like dizziness; the fate of the world swinging wildly back and forth, while my own life got meted out in days, one after another, effectively unchanged. Here was solid concrete, warm weather, a white steeple. The oblivious rectitude of the moment. I wanted in that instant to tear it all to shreds, to get out of it for even a second.

I took out my phone and decided to call my father. Since high school, we'd begun to talk more openly about the Problem. It was no longer just a subject of his research, it was all over the news—and this somehow made him *more* capable of talking about it, as if mounting public alarm lent him a framework through which to feel the facts he'd helped produce.

He picked up on the first ring, as he often does when I call. He was angry he told me, *livid*, though his voice sounded like this was something he was still trying to muster. (I loved him so much then, the way he sought to transcend his natural gentleness, will himself into indignation; loved him because he tried, loved him because he couldn't.)

I wish I could feel angry, I told him. All I feel right now is sad. And then, like I'd issued a summons, the tears arrived and I was not in control of them anymore, they were just falling down my face, wetting the screen of the phone. Along one fence of the parking lot was a flowering hydrangea bush, and I began plucking its leaves and blowing my nose into them, dabbing my eyes with a sleeve. For a moment I felt like I was eight years old again, crying to my father in disbelief; like I was reliving that conversation we'd perhaps never had.

But even as I experienced it—even as the force of it made me sit down and slump against the fence—a part of me was already bored with my grief. It felt repetitive and dull, an exact replay of my reaction to *Melancholia* and to Hurricane Sandy and to all the other moments when the weight of the hyperobject had ruptured the strength of my resolve. It was like the Problem had placed my emotions on an endless tape loop, and they were going to keep playing back at me forever, in the same notes, the same sequence.

With this feeling came the equally predictable guilt of having never occupied a position of real vulnerability to the Problem—a position that would warrant true sadness, like if your father killed himself after another crop failure, or a hurricane shredded the house where you'd grown up. The guilt exhausted me, doubly so because I suspected its actual purpose was to exorcise the dissonance of privilege by effecting a moral catharsis that would mean absolutely nothing to anyone beyond myself. Sitting there in the church parking lot, it was like my whole life was being stretched out in front of me, just one long sinusoid of elation and despair, an infinite rerun of whatever I'd already felt.

This reverie was interrupted by the voice of your grandfather, who was saying something I couldn't quite hear. "What?" I asked him, stifling my tears.

"Maybe this will have the opposite effect," he repeated into the phone, a little louder this time. "Maybe all the other countries will rally around it now and make the accord even stronger. I think you should still have hope."

"I do have hope," I told him. And it was true. I felt it like a sliver in my side, pulsing there, bating my breath in defiance of my brain. Sometimes I wished I could extract it and let the wound bleed and heal until I didn't feel anything there anymore, didn't have anything I had to attend to. But hope possessed me with a terrible vigilance. Even after the worst news, it made my heart bounce back like a reflex, no less exhausting that it was automatic.

I do have hope, I said again, and we both sat silently there on the phone, like we were waiting to see if it would hold.

Walking back from the church my tears dried up and my nose cleared. I felt like I was watching myself from afar, the distance from my own life telescoping between the sidewalk and the stratosphere, my mood oscillating with the frame.

I tried my best to let it change, to move with it. Better that than the armored solipsism I knew myself capable of: *all that's happening right now is that I'm walking down a side street in Buffalo.* No, better to let it all in, let the Problem suffuse the world and the world suffuse my body, let it flood through me and spill over the sides.

It was June and the sun was shining hot on my shoulders. I pictured the tiny photons like dust motes, trillions of them glancing off of me, sifting through the weave of my shirt and falling into the forest of my hair. They were almost nothing, I knew, nearly massless, plinking into everything always. And yet this was the crux, this was what we were all litigating—the sunlight itself, and how much of it would be trapped here with us.

I shouldn't have, but I felt then for a moment like it was all very simple. Beneath the text of the treaties and the crack of the gavel and the ever-bobbing derricks there was just a density of sunlight, a single question. It was all so clear: the Problem could be seen, was in fact the *means by which* we saw.

And from out of this thought came hope, seeping back into me, unbidden and indomitable. Or maybe that's not where it came from. Maybe it was just a beautiful day, and I was out for a walk in a quiet neighborhood. Maybe it was just spring. The thing is, I never quite knew its provenance. It would just always come back, offering its hand, saying trust me. And once again I'd take it, feeling tired but a little thrilled. It came back, I'd think. Even after all this, it came back.

PART II

PART II

Retreat

Back when I first started the job with the NY Renews coalition, I made the decision to begin seeing a therapist. In our initial consultation, I tried as best I could to articulate the source of my anxiety. I had a sense, I told him, that the Problem was pulling me into the future against my will, and that this pull was constantly accelerating.

This was true first of all in the obvious sense. My work for the coalition was focused on accomplishments decades in the future, goals we were trying to codify through legislation: 50 percent renewable energy by the year 2030; 100 percent renewable energy by 2050; all anthropogenic emissions eliminated by mid-century. Across the globe, governments and nonprofits were setting these kinds of far-off goals, unveiling them at press conferences with words like "launch" and "mission," as if it were still the 1950s and they were announcing their plans to put a man on the moon. This parallel was often made, in fact, though there was no pretending it was the same thing. These new goals were not simply challenges we'd set for ourselves out of curiosity or ambition. They were deadlines, effectively: due dates prescribed for us by the increasingly alarming evidence being pulled out of our atmosphere. And as deadlines do, they hung over our heads, growing more urgent and less plausible the longer they were put off.

But the dynamic was also true in the more prosaic sense of needing to plan what to do in the meantime. Much of my time was diced into half-hour increments—meetings, calls, actions, all of them trained toward the far-off deadlines. Most mornings this was the first thing I looked at, my day portioned out in neat purple blocks on the screen of my phone. I could zoom out once and see the full week; twice, the whole month, a checkerboard of purple and white. By my early twenties, work on the Problem had filled my calendar with deadlines on the hourly, daily, monthly, and decadal scales. There was still some ambiguity left on the level of years, though in my mind this was getting increasingly constricted by those time-sensitive events I associated with life's second quarter: grad school, spouse, house—and you, possibly. Meanwhile, appointments were wrapping up or running late, new ones juggled and booked. Time felt cramped this way. It had no give. The hours passed bumper to bumper like traffic.

This was one of the major topics of conversation at my standing appointments with the therapist, which began at 8:15 a.m. and ran for forty-five minutes. Every Friday, I'd take the train to his office on the Upper West Side, a cuboid room on the second floor of an expensive-looking brownstone. The room was neatly arranged, with two red armchairs facing each other across a Turkish carpet. Placed discreetly next to my armchair was a box of tissues, which I never once used. (Though I often felt like I wanted to, I could never will myself to cry during therapy.) The walls were hung with paintings of harbors and lighthouses, and, incongruously, the bleached skull of some impressively horned ungulate.

"It's like you're in search of lost time," he told me in one of our first sessions. "Except in this case, the time you're searching for is all out ahead of you." He liked to make these kinds of literary allusions, though I suspected this was not a pretension he adopted with all of his clients but rather a concession to my own, an appeal to common ground. At the time I'd never read Proust, though I nodded knowingly in agreement. "I think you're right," I said, not sure yet if I really did. "Maybe my head is so stuck in prospective mode that I'm having trouble living in the present."

At the end of the session he pulled out a slim volume from his bookshelf and handed it to me. It was called *The Power of Now*, and the dust jacket had a bad font overlaid on an unseemly yellow-to-green ombré. I accepted the book a little skeptically and spent the next week reading it on the subway, as the trains lurched from home to work and back.

The book had been written by an elfin-looking German man of no stated religious affiliation named Eckhart Tolle. Its main thrust seemed to rest on the claim that the past and the future were not in any important sense real. Dwelling in the former and worrying about the latter were the chief sources of suffering for the human ego. The trick, apparently, was to hitch your mind to the present and ride with it on its twisting path, accepting whatever came, up to and including being sardined on the 6 train as the delays made you late to a meeting that was itself decades delayed in its attempt to address the Problem. If you did this right, I learned, it was called "Living in the Now."

My favorite example of Living in the Now had to do with ducks. In one part of the book, Tolle asks you to imagine two of them gliding along perpendicular paths in a pond. When the paths meet and the ducks collide, there ensues a momentary kerfuffle of honking and wing-flapping. Then the ducks disengage and swim inscrutably onward, as if nothing has happened at all.

This anecdote was meant to demonstrate the serene bliss of the Now: a human mind might have spent time smarting over the spat or anticipating the next one, but the ducks gave it no more or less time than it took to elapse. Perhaps if I could imitate the ducks and take each moment as it came—as neither a projection of the past nor a prelude to the future—then the Problem would loosen its grip and I'd be able to live a little more slowly, at the pace of my days, which, despite their impression of relentless acceleration, were still technically going by at a rate of one every twenty-four hours.

My first attempts at locating the Now were unsuccessful. One Sunday, soon after reading the book, I found myself with an afternoon off and decided to bike out to the usually quiet and depopulated Woodlawn cemetery, in the north Bronx. I got a later start than I wanted—having spent the morning pruning my inbox—and when I pulled up to the gates the shadows of the marble headstones were already lengthening, slightly offset, like the black keys on a piano. It was late winter, and I had to crunch through melting snow with my pannier to find a place to sit. The cemetery had the feeling of a foreclosed subdivision, the sidewalks abandoned and the mausoleums padlocked. According to a sign at the entrance, Herman Melville was buried in one of them, though I was unable to locate it among the twisting paths and eventually settled down on a snowless patch of grass on an arbitrary hill in what I took to be the middle of the cemetery. The ground there was still wet with meltwater, and as soon as I sat down it bled through the blanket I'd brought and soaked the leg of my pants. I got up and moved uphill into the lee of an obelisk, scraping away the snow with my boot and sitting down again with my back to the granite. But when I tried closing my eyes the cold was sharp and angled, and I kept having to fix the blanket to avoid sitting on the wet parts.

Amid these small adjustments and procrastinations, the Now evaded me, and my mind wandered back to Melville, to what I remembered of the particular chapter in *Moby-Dick* in which he recounts the experience of an anonymous deckhand up in the crow's nest of the *Pequod*. The young man is meant to be watching for the spout of the white whale, but his gaze keeps drifting back to the horizon, the circumference of the world, which on the open sea never changes. Even as the ship plows through the surf, he has the impression of being stationary, alone at the center of an endless plain. He turns his cheek to the sun, lets his lids droop closed. And though the crew is butchering their catch down below—cleaving meat from blubber, bone from bone—he cannot hear them at all; their noise has been lost to the wind and the height of the mast.

Maybe, I thought, this was something I could take back to my therapist, another pretension for us to riff on. I would tell him that it captured what I most envied and feared about life in the Now. How you could acquaint yourself with deep peace, but only by ignoring the carnage.

Sitting there beneath my obelisk, I scanned the cemetery for any boundaries that might keep the former in and the latter out: the far-off thickets of pine, the line where the snow met the pavement, the muffled peal of traffic. Nothing seemed to hold. I had no vantage, no breeze. I tried closing my eyes again, but my thoughts sped backward and forward through time, dragging me along in their wake. I pictured the municipal flood maps, a habit by now—how Woodlawn was going to be fine, elevated safely above the worst-case scenario. I pictured the last glacial maximum, how massive ice sheets had ground to a halt at exactly this latitude, their white flanks looming high over the ancient Bronx. I pictured the calendar on my phone, where I had blocked out a meeting for eight o'clock the following morning. And then all of a sudden I had to leave, packing up my things in a hurry, folding the sodden blanket into itself. I got home that night with just enough time to make a big pot of stir-fry, which I doled out into Tupperware containers and stuck in the fridge, one for each day of the oncoming week.

What I think I was feeling was this: that time was both too big and too small. That it was being stretched longer and sliced thinner than I could possibly conceive.

On the one hand, the Problem now regularly prompted me to think out past the bounds of my own lifetime, way out in both directions, the aperture of my involuntary attention widening to include events that had preceded my birth by thousands of years and would succeed my death by the same margin. But on the other hand, its urgency seemed to be dicing my days into smaller and smaller increments—half-hour phone calls, fifteen-minute task windows, a breathless acceleration built of compounding deadlines. In this way, the past and future loomed ever larger while the present grew clipped and pressurized. Little by little, time itself came to seem like a strange kind of resource—pervasive but scarce, effectively limitless but essentially nonrenewable.

———

IN THE WEEKS AFTER the United States' withdrawal from the Paris accord, everyone in the coalition tried their best to shore up

morale. We cannot retreat, we repeated to ourselves and one another on conference calls. Retreat is the last thing we should do.

Months later, I received word that I'd been accepted to a
writers' retreat in northern California. I'd applied guiltily, knowing this was exactly what I wasn't supposed to do. But I was beginning to feel desperate for time. The type of time that could
still seem spacious and relenting, that wouldn't require I translate my grief straight back into work. If I kept on doing this, I
worried, I'd soon wear through them both and withdraw into
apathy. So although it felt like an indefensible indulgence, I applied for a few weeks' respite, hoping it would feel like much more.

By then, I had been writing to you for months, and wanted
to craft something coherent out of the scattered notes on my
phone. Prior to the Paris withdrawal, I'd still been able to entertain long-shot fantasies that the trendlines might be reversed
before you grew old enough to read them. After the withdrawal,
these fantasies became harder to sustain. It felt increasingly
clear that the Problem would be there, waiting for you, worse
than it was now, and that I would need to find a way to talk to
you about it. Not only talk, but advise, or console, or hand you
something with which to face the hyperobject. This seemed suddenly like an intergenerational responsibility: to see whether
I could defensibly tell you what my father had told me—that I
think you should still have hope.

When I applied to the writers' retreat, I focused on my trouble
with time, though I groomed my anxiety so it sounded more like
an authorial conceit than a full-blown existential crisis. In my
pitch I developed a new language for it. I set aside concepts of
past and future, tabled tense entirely. What I'm really after, I

told them, is a bridge between time*scales*, this feeling of being alive in both biographical and geological time, of living with and through their disjuncture. A sense that I could see the twig of my lifeline balanced tangentially on the far longer but still briefly coincident vector of the planet itself, and how it was scary but I wasn't scared. How I somehow couldn't be. How I was still sending emails, and getting haircuts, and pouring milk over my cornflakes like everything was normal, which it often deceptively felt like it still was.

———

THE RETREAT WAS HELD IN SPRING, in a beautiful house on the coast of California. When I arrived, the residency director gave me a tour of the grounds, which were elaborate and thick with flowers. Glossy California poppies hemmed the garden path, and Indian paintbrush stuck straight up from the bushes. Inside, the house was hung with tasteful artwork and writerly quotes. "When the heart speaks, take good notes," said the one outside the door to my room.

There were three residents at the retreat, and each had been given a small writing shed in back of the main house. The director led me down to mine and we both stooped through the low door into a room furnished with a rug, a desk, and a small sofa. The inside smelled like cedar and the front window looked down on a salt marsh, which was swollen with high tide.

"Under there," the director said, pointing with some drama at the marsh, "lies the San Andreas Fault." They didn't publicize this on the website, she explained, for fear it would scare people away. Hearing this, I felt the opposite of dissuaded—I couldn't believe my luck. I looked up and down the marsh and it seemed that my little hut was the closest physical structure to the exact location where the Pacific Plate was grinding slowly past the North American at a rate of two inches every year. In the most literal sense, I had stumbled upon the boundary between biographical and geological time.

My first day at the retreat, I could barely work, I just watched the tide furl back across the marsh, exposing a maze of muddy channels and tufts. The fault line offered almost no indication of its presence, aside from the marsh itself, a subtle dip in the land where the water from the bay had flooded what was once a cow pasture. I waited with my bare feet on the floor and my fingers poised on the keys, hoping to sense faint tremors from below, some ripple beneath the present. But the San Andreas stayed quiet and the clock ticked out its rhythm and eventually I shut my laptop, having written next to nothing in biographical time.

My two co-residents were friendly but reclusive, and the retreat turned out to be pretty lonely, a feeling I'd grown unused to in the world of conference calls, and which I soon found to contain all sorts of inner shades of boredom, melancholy, distraction, and joy. In the afternoons when I'd finished a spurt of writing, I would take walks on the nearby interpretive trails, keeping my own company by reading all the vinyl-coated informational placards. On one trail, I read a placard that described the process behind earthquakes, how the tension along the boundary built up over decades and decades until the rocks cracked and the plates shot past each other. The last time this had happened, one of the plates along the San Andreas had jumped twenty feet in one second, though I couldn't make out whether this had been the Pacific or the North American because part of the vinyl had bleached and curled in the sun. Farther down the path, a bisected fence had been built to visualize the magnitude of this leap, one half sitting twenty feet in front of the other. I imagined my shed sliding twenty feet north relative to the salt marsh—how similar the view would be, despite everything having changed in an instant.

Here is another thing I learned: that the machine used to measure earthquakes is called a seismograph. It uses springs and weights to record each tremor onto a sheet of paper, creating a scribbled graph that can be read in real time. Even from hundreds of miles away, a seismograph can tell you the exact magnitude of a catastrophe, transform it into reams of data on a printout. But that still isn't the same as feeling the earth shake.

That night I felt too anxious to sleep. Though I'd actively applied for the residency, I somehow hadn't prepared myself to transition from fighting the Problem to having to think about it. I lay in bed, waiting for the thoughts to come, so I could turn them into the work I was meant to produce. This felt like just another task, something that could be squeezed into a block of time. I could see the block, purple and capacious on my calendar, its ending already in sight. But though I willed my brain to move, it sat there paralyzed inside my head.

Outside, the flowers were all loosing their pollen. My throat itched terribly, and I kept rubbing my tongue over the roof of my mouth, making dull suctioning sounds. At midnight, I threw the covers off and lurched to the bathroom where I popped a small white allergy pill directly from its foil pouch into my mouth, realizing only afterward in the dim glare of the wall-socket night-light that the box read "Non-drowsy, 24 hours." I did not know whether non-drowsy implied the presence of a

stimulant or merely the absence of a soporific, so I went back
to bed, waiting for my brain to indicate one way or another. My
eyes stayed open and reluctantly adjusted to the dark, revealing
the gray edges of things, the corrugated glow through the blinds.
The crickets outside were rubbing their legs into a racket. I tried
putting on my sleep mask, staring sightlessly into its black satin,
or closing my eyes and watching colored smudges bob across
the insides of my eyelids. But the anxiety persisted, as did the
allergies. At 4 a.m. I admitted defeat and stumbled once more
to the bathroom, where I took a small pink Benadryl, feeling it
slowly turn the lights off in my head, like it was going room to
room in a house.

I woke up at 11 a.m. in a pharmacological haze and dragged
myself to the shed. My eyeballs felt heavy, like they'd been replaced with glass eyes, the big realistic marbles you have to pop
into the socket and adjust with your fingers. I stared into this word
document, hopelessly drowsy and impossibly awake. There was
still no movement from the fault, though out on the salt marsh
I spotted a family of ducks gliding single file up a tidal creek.
The residency had provided me with a pair of binoculars, and I
lifted these to my glass eyes and watched until the last of the
birds crossed the San Andreas and hopped out onto the mud of
the Pacific Plate, disappearing from view.

To fall asleep the next night, I returned to my old strategy of
picturing the whole world flooding inch by inch. I started with
my bed, how the water worked its way up the wooden bedposts,
soaking into the mattress like a sponge, flooding into my nose
and mouth and then gently lifting the sheet off of my body so
that it hovered above me like a ghost. The water rose in the salt
marsh, erasing the patterns it had carved in the mud, rising as

one solid plain until it brimmed over the banks and into the flower beds, stands of lilac and paintbrush buoyed like seaweed in the current. Then it reached the writing shed, seeping underneath the door and lapping at the legs of the chair where I sat cross-legged, trying to finish a sentence before the laptop drowned and went dark, words blinking out on the screen, blotted from the notebooks on the desk. And it kept going, swallowing the peaks of things, the roof of the shed and the tops of the pine trees. The needles came loose from their branches and schooled upward like fish toward the crests of the hills, which for a moment were still just visible beneath the surface, folding the water into waves, until they too were submerged and the tide finally closed over into an undifferentiated Pacific. From below, it was like looking up toward the roof of a vast stadium, empty space imbued suddenly with volume, the surface impossibly high above. Bubbles rose in towers from the sunken marsh, whales shuttled past in the gloom. The whole world felt dark and buoyant. I sank into sleep, drowned in it.

————

BACK IN NEW YORK, a few months before coming to the retreat, I'd had an appointment with my therapist the day before my twenty-sixth birthday. I remember telling him—with a vehemence that surprised us both—that I hated 1990, the year of my birth. How my generational cohort had had the bad luck of being born right when the Problem was first emerging into public consciousness, how we'd had to watch over the next three decades as it transformed from McKibben's urgent threat to Kolbert's looming inevitability, how this whole process had tracked our coming of age. Better to have been born in ignorance a century earlier, I told him—or even a century hence, when things might really start to deteriorate. Anything but this awful century of waiting, of *bracing*.

My therapist smiled when I said this, seemed to fish around for something. "Perhaps you're an anti-natalist," he said finally, half musing and half asking. "Perhaps you believe it would be better not to have been born at all."

Put this way it sounded ridiculous. By all reasonable benchmarks I had an exceptionally good life—nothing about the Problem changed the fact that being born white, male, and middle class in the Global North on the eve of the new millennium was like winning a lottery of ease and opportunity. There I was, after all, paying ninety-five dollars an hour to sit in a comfortable chair on the Upper West Side and talk through my feelings under the gentle guidance of a professional.

Still, when I got home that night, I spent hours online reading about anti-natalism. After a while, I found myself scrolling through the website of something called the "Voluntary Human Extinction Movement," which, according to its mission statement, advocated for "the cessation of all human reproduction so that the biosphere can return to full health." On the home page was a cartoon of a person waving and smiling alongside a dinosaur and a dodo bird. Below them read the motto: "May we live long and die out." (The list of FAQs was equally deadpan: "Are you really serious?" "Do we have to stop having sex?" "Are some people opposed to the VHEM concept?")

I clicked through to a video clip of the group's founder, an unassuming man named Les Knight, doing an interview with Tucker Carlson. Knight spoke softly and was dressed mostly in khaki. He looked like a mailman or a philatelist. With surprising good humor, he was telling Carlson about his decision to get a vasectomy at twenty-five, arguing that in an ideal world everyone would make this same choice. That way, he said, humanity could live out its last generations in peace and prosperity, our numbers gradually diminishing until, eventually, the last of us would cede the stage back to the thousands of species we'd driven to the brink of oblivion.

"Why would you do that," Carlson asked, "why would you want people to become extinct?"

"Well it's either us or a million other species," Knight said with a chuckle.

"Yeah, but I guess the obvious answer is, what if we prefer our species to other species? Isn't it fair for human beings to want to perpetuate their own species?"

"Well it would be fair if that's all we did, if we also let the others survive and exist. But ever since we became Homo sapiens, we haven't been able to do that, we've sparked a mass extinction."

"What do you mean? There are many species of plants and animals that are thriving right now. There are all sorts of, um, insects and algae . . ."

Knight's logic seemed unassailable, but I couldn't help it—for the first time in my life I was on Tucker Carlson's side. I sputtered along with him, ablaze with anthropartisanship. What about our art and our invention, I justified to no one; what about our as-far-as-we-know-unique-in-the-universe capacity for self-reflection?

But it was too late, I realized: the argument had gotten into me, was already scattering its seeds. It would sprout later as a question, urgent and unanswerable, rampant as a weed.

For some time after that, I developed a habit of imagining I had a dark, anti-natalist twin. He would appear to me vividly, like a reflection in the mirror, popping into my head in the most ordinary situations. I visualized him at meetings, sitting in the back of the room, lofting unhelpful comments and distracting side-bars, trying in his furtive way to sabotage any work on the Problem. Eventually he'd be cornered, forced to admit his motives. I pictured him coming clean in a ranting, defiant soliloquy. Imagine, he'd say, if Homo sapiens survived long enough to escape planet Earth and colonize the galaxy. Imagine if this cancer species were allowed to spread!

I know I am not him, that I do not agree with him, but, as with any evil twin, there's an unsettling familiarity there, a kind of magnetism. I can tell you that we differ, but I cannot tell you exactly why.

Even as I began writing this letter, there was a small part of me that thought it was pointless, that hoped you would never exist to receive it. Every morning of the retreat I'd shuffle out to my little hut, nervous somehow, though there was nothing riding on any of it—few people even knew I was there. I kept a stack of books next to the binoculars on my desk, and I would glance through both of these whenever I felt stuck, which was often.

At the top of the stack was Maggie Nelson's *The Argonauts*, a book that I love and return to every time I need reminding of what agility means, what care and courage look like in prose. It too revolves around parenthood, in its way. Nelson chronicles the formation of her family: falling in love with her gender-fluid partner, learning to care for his son, trying and failing and trying again to get pregnant before finally giving birth to a child of her own, bringing him into their tender little fold. When I first read the book, I took serious heart: here, I thought, was a love that might survive the Problem—a love anchored in change, defiant of category; a love whose parts could be overhauled one by one without compromising its integrity, a love that was in fact *built around* this kind of endless transformation. It was from *The Argonauts* that I learned what it might look like to write you without defining you, to love you fiercely without turning away from all the complexity that love implies.

And yet, flipping through the book one morning in my shed, I realized that even Nelson would probably have had doubts about the letter I was struggling to write. I found them in a passage I'd forgotten about (though must have noted at some point, since the page was sharply dog-eared):

> I've always been a little spooked by texts addressed to or dedicated to babies, be they unborn or infant. Such gestures are undoubtedly born from love, I know. But the illiteracy of the addressee—not to mention the temporal gap between the moment of the address and that at which the child will have grown into enough of an adult to receive it . . . underscores the discomfiting fact that relation can never be achieved in a simple fashion through writing, if it can be achieved at all. It frightens me to involve a tiny human being in this difficulty, this misfiring, from the start.

I feel sad reading this now, embarrassed by its rightness. As if the Problem weren't enough for you to deal with, as if on top of that you needed the weight of your father's grappling, all my clumsy attempts at connection. My instinct is to apologize for handing you this burden, though this would of course assume that you're a foregone conclusion. And thinking through Nelson's critique, I'm reminded that you are not, that no child ever is. Even tabling the possibility that my misgivings might get the best of me, or that something else could go wrong, medically or financially, there remains the basic fact that the decision to have you will be just as much my partner's as my own. Which is simply another way of expressing that, while your reading this would indeed imply your existence, my writing it does not, necessarily.

I'm coming to realize that my letter is a hedge against all this uncertainty. If building a family proves not to be an option— that is, if conviction or circumstance ultimately precludes it— then writing to you may be my only opportunity, however pathetic and imaginary, to feel something along the lines of what Nelson calls relation. In which case, a letter to you really just becomes a letter to me, replete with its own misfirings, its own blend of hurt and of care.

Reading Nelson raises another doubt for me, too: that the group I most associate with the anthropomorphization of unborn children—the group that most frequently indulges in this same mode of address—is the anti-abortion lobby. One of the big ironies of the anti-abortion cause is that most of its adherents, for all their hand-wringing about foreclosing the future of the unborn, have chosen to ignore a Problem that's doing precisely that, albeit on a universal, planetary scale. Given their genuine concern about protecting potential children, shouldn't they also want to ensure those children inherit a world capable of sustaining them?

Perhaps here is where they'd accuse me of hypocrisy: that I, too, am fetishizing the unborn in an attempt to shape the policies of the living. But I am not writing to the unborn. I am writing to you: my own specific child, the only one over whom I'll ever have a say. And my aim here is not to wield you as a political cudgel—like I would ever be so naive as to imagine that a heartfelt letter could convince the Pruitts. I just can't see how I could decide to have you without first trying to humanize you, to address you as a person. This whole letter is premised on the fact that this is an intensely personal question, and that it's within my power to deliberate and choose. Denying that same choice to women has always been immoral, but it seems particularly sadistic in the era of the Problem: that the Pruitts would set the world alight and make it illegal for a woman to choose *not* to bring a child into the blaze. Still, I feel a small trace of empathy for the opponents of abortion. At least, I can understand loving someone who exists only in the hypothetical, how strange and fraught it can feel to enact that love in the actual.

On our last night, the residency provided us with a gift card to a local restaurant, and my co-residents and I emerged from our sheds to walk to town for a meal. It was late afternoon and the light was painting the salt marsh gold, summoning the crickets from their holes. We passed a few old barns, a horse in a paddock. I was happy to be out of my chair, away from the fault line.

At the restaurant we scanned the menu of locally caught trout and sustainably raised beef and placed our orders. Conversation turned predictably to our work. One of the residents spoke about the play he was drafting, which was set in Louisiana gulf country after the BP oil spill. The other said she was writing a book of queer science fiction set in a dystopic future India. I went last, wishing we could just keep talking about their projects, which sounded more interesting anyway. I told them briefly about my organizing work with NY Renews, and how I was doing some writing to process my feelings about the Problem. (I did not tell them about you.)

"How do you keep doing all that work fighting the Problem,

given the predictions?" asked the novelist. She said this in a tone that I think was meant to sound laudatory, but just served to make me feel tired. I issued my standard reply about how the Problem isn't binary like nuclear war: it's not like we either press the button or we don't. Even if it's ongoing and inevitable, there's still a world of difference between two degrees Celsius and six degrees Celsius in terms of human suffering and general chaos, and so every marginal bit of good we do in the present allays some of that pain in the future. "If that makes sense," I added, my usual caveat. "No, no, that's very rational," said the novelist, looking back over her shoulder at the waitress, who'd returned with more of the complimentary popovers.

It wasn't untrue, but there was more I should have said that night: that I didn't have a perfect answer—that I wasn't always sure how to keep working on the Problem, didn't even know if I still was. What about this letter? I should have asked. Does it count, in your opinion? Is this what we mean by work?

———

ON SEPTEMBER 20, 2017, at a theater in Berlin, the French theorist Bruno Latour gave what he called a "performance lecture" about the Problem. The performance consisted of him, onstage in a rumpled jacket and tie, having uncaptioned images projected over his body and onto the large screen behind him. Latour is a proponent of the idea that all knowledge is socially produced, that truth does not exist apart from the entities—both human and non—that have collaborated in its "discovery." Watching the lecture, the projections seem intended as a reminder of this conviction—how the images representing his ideas are draped over his physical frame, which appears hunched and diminutive at the bottom of the screen.

At first, a giant, rotating image of the planet appears in the dark. Small pieces of ocean and continent pass over the topography of his face, its prominent nose and jutting eyebrows. Then the screen switches to a picture of black and white geologic strata, and he raises his arms in emphasis, layers of rock pinstriping his sleeves.

To Latour, the advent of the Problem means that the whole world is now socially produced, that there is no "outside" from which to view it. He tells his audience about a flight he took that passed over the ice caps of Baffin Island, in remote northern Canada. Looking out the window of the plane, he could see cracks and ponds in the ice, evidence of the accelerating melt. He caught sight of one large sheet of ice that had melted into a shape that looked to him like Edward Munch's *The Scream*, with two small pools for the eyes, and another distending into an anguished-looking mouth. "It seemed like the ice was actually sending me a message," he says.

Before, when you witnessed a spectacle of nature like this, it was possible to witness it from the outside, merely as an observer. But now, he explains, the plane he took was having an effect on the spectacle he witnessed from its window, could in fact be said to have helped produce it. We've cornered ourselves, in other words, in a Heisenbergian trap, where our activity as observers inevitably impacts the things we observe, where we've managed to catch all of nature in the growing dragnet of our social determination. For this reason, he explains, he has titled the lecture "Inside," the name of what is now the only relevant place, the place where we all live.

Onstage, a photograph of the face that emerged from the ice

emerges now on Latour's own face, the projection lending him the bright white pallor of a ghost. It does look like *The Scream*, though only slightly more than might be attributed to the power of suggestion. Then the screen flips back to the image of the rotating globe.

In this new age of inside, he argues, images like this are no longer useful: a blue-green planet seen from the outside, shot backward through the shuttle window. This is the globe envisioned by globalization, an ample orb revolving in space, big enough for capitalism to bestow its hypertrophied strain of prosperity on all nine billion human occupants. But this place does not exist. It would take many Earths to accommodate the grand telos of global capital, and we've already almost used up our one, distributing too many of its fruits to the Global North, denying the Global South even its most basic needs. For Latour, then, the image of the globe is just an abstraction. It is not where we live, nor where we should imagine ourselves as living.

The place where we actually live is in what scientists call "the critical zone," the thin pellicle of crust and sky where everything we've ever cared about has happened—all of history, of evolution itself, contained in a few vertical miles. This is our real home, Latour says: not an ample orb but a fragile skin, the fuzz on an indifferent peach. And to him, it is less a place even than a process, what he calls a "vortex" of nested cycles: blood and food through the body, water and carbon through the earth. Minutes aswirl inside days inside decades inside eons. It takes humility to admit that we're inside this process, that we are implicated, that we cannot step out and observe its changes at a distance. To live with the Problem, Latour argues, you have to stop trying to read the world from above. Read it, he says, "as if you're on the surface of the parchment."

I first watched this lecture on the promise that it might hold some answers to the Problem, or at least satisfy my desire to witness someone who did. But though I envied his clarity and conviction, there was something abstruse in Latour's conclusions. Even now, as I set them down on the surface of the parchment, his words are difficult to internalize. I can understand what they mean, but I'm not sure what it would *feel like* to inhabit the worldview they describe.

Still, that first time I was introduced to them, Latour's methods did make a certain conceptual sense: understanding a vulnerable planet by examining its thin skin, measuring "the reaction of the skin to its own activity." This sounded not unlike the daily meditation practice I'd taken up around that time as part of my therapist-recommended quest for the Now.

Every morning I would set the timer for fifteen minutes and sit on a small red cushion I'd placed in front of a window, scanning my attention up and down the surface of my skin. The goal was to stay with whatever sensations came up, trying not to picture my body from any representational remove but instead just allowing myself to experience all the little vicissitudes that made it tingle and cramp. Because the window was at street level I kept the blinds closed, since the thought of someone seeing me as they walked by—sitting still with my eyes closed, unaware of whether and by whom I was being seen—made me feel vulnerable, like I was seated in front of a one-way mirror. The few times I had left the blinds open, I'd been distracted by this outside image of myself, a motionless man witnessed by a stranger at a distance. So instead I would stare into the folds on the back of the blinds—white hexagons made of some sort of delicate crepe paper—and ring a resonant metal bowl that I kept on a low shelf next to my cushion, waiting till the chime waned asymptomatically into silence before finally closing my eyes.

Some days the meditation felt good, like I was building a quiet place in my mind impervious to the tremors of the Problem. My therapist called this the "inner shrine," though privately I'd begun to think of it more like a cognitive version of the doomsday bunkers outside of Topeka: a silent redoubt buried deep in my head, somewhere I knew I'd be able to go if things got really bad.

Other days, the meditation felt impossible, like I was trying to stop time, to pluck a ripple from its current and freeze it solid. I would sit there in anguish, trying to grasp the Now, to prevent it slipping through my fingers, though inevitably it would anyway, coursing forward toward the Problem like a river toward a cataract. On days like this, I would sometimes postpone the moment when I'd have to open my eyes, knowing that when I did I'd be looking out on the future, which would have once again effected its constant and relentless arrival.

———

A FEW MONTHS AFTER RETURNING from the writing retreat, I embarked on another sort of retreat, this one to a meditation center in Massachusetts. At my therapist's suggestion, I'd blocked out enough room in my calendar for a ten-day course, in hopes I might finally locate the Now myself and then write you a map on how to find it.

On the drive up, the forests along the interstate were still mostly green, though I couldn't remember whether this was normal for late October, whether the foliage was supposed to have changed already. I arrived at dusk to a cluster of low, prefab buildings tucked into the woods off a rural state road. On top of one of the buildings was a large stupa, its filigreed bells silvering in the wind, the only devotional flourish in the otherwise functional-looking compound.

At registration, I handed my phone, keys, and wallet over to a volunteer, and signed an agreement stating that I would abide by the guidelines of the retreat: no talking, no reading or writing, no sexual activity of any kind, and no exercise more intense than short walks or simple stretching. I was taken to my room, a sparsely furnished double that I was to share with a tall blond man about my age. We were brief and noncommittal in our introductions, knowing that after dinner we'd begin our silence and wouldn't be able to communicate for the remainder of the ten days.

The schedule of the retreat was precise. We were woken at 4 a.m. every day by the sound of a gong struck several times in the hallway of the men's dormitory. Ten minutes later we would all pile out of our rooms and shuffle en masse toward the meditation hall, wearing loose pants and slippers, some of us with shawls wrapped around our shoulders. There was something comforting and unusual about the company of grown men who couldn't talk, everyone groggy and tender-looking in their pajama bottoms, no one joking or jockeying or even making eye contact. In the hours before sunrise, the hallway had the chill, interior glow of a hospital night shift. All you could hear was our padded feet on the linoleum.

The meditation hall itself was a large, bare room, dimly lit and paneled in wood. Participants sat on blue cushions arranged in a grid on the floor, and there was a low dais at the front for the teachers. On the first day, their instructions were simply to observe our own breaths as they went in and out, resisting the temptation to count them, or visualize them, or furnish them with any mnemonic whatsoever beyond what it actually felt like as they passed through the nostrils. Over the course of the day we did this for ten hours. For the first few sessions, I could barely sustain my attention for a single minute. It kept springing leaks, punctured over and over again by my brain's attempts to gauge its inviolacy, like I was feeling around for holes with the tip of a knife. Even thinking about avoiding distraction opened up a crack through which could squeeze the names of a few things I wanted to avoid being distracted by, followed against my will by their images, which of course widened the crack even further until the whole tentacular mass of their possible associations flopped unceremoniously on deck, writhing around and upstaging my breath until I summoned the presence of mind to intervene, refocusing my attention on its muted little putter.

Still, I learned some things. I was embarrassed to discover that my two most common distractions were, on the one hand, a simple, free-floating lust, and on the other, a burning rage against the Pruitts—both in their way related to you, to my desire to see you live. I learned that on the way in my breath felt like a cold plume against the back of my throat, and on the way out it was warm and plump inside my nose.

When a session ended the gong was struck once more. Everyone gathered their shawls and shuffled back down the hall toward the dining room, where we were served two simple meals a day. In the middle there were a few long tables set with metal folding chairs, but I preferred to sit in one of the chairs facing the wall, looking at no one and listening to the sullen chorus of cutlery on dishware.

In the short periods between meals and sittings, we were allowed to take walks on the center's modest grounds, a small network of dirt paths surrounding the buildings and dipping into the woods. When everyone was out it could feel almost crowded: dozens of men clasping their hands behind their backs, making subtle adjustments in trajectory to avoid running into one another. During these breaks, I liked to sit on a white plastic chair that had been placed on a bed of moss next to a small stream. I would try to watch a single bubble or twig as it rode the current, tracking it from where it entered my view at a slight bend upstream until it disappeared in a minor falls farther down. Usually I didn't get very far, losing it after a few seconds.

During these first few days, I often returned to my room to find my roommate lying in bed with the covers pulled up to his chin, holding his stopwatch in front of his face and examining it with an intensity bordering on anger. On the third day, I came back from an afternoon sitting and found that he'd left, taking all his things with him and neatly making his bed, so that there was no trace of him ever having been there. Though I racked my brain, I couldn't remember what his name was, if I'd ever even learned it before the silence started.

Say what you will about the Buddhist belief in reincarnation, at least it does not rush you, hold a stopwatch to your face. In the secular world, time can be kept, wasted, spent, and used wisely, but it can never be turned back. So even if I don't return as a fish or a fox, I can see the wisdom in imagining that I might: there is a resilience, I think, in seeing time as a renewable resource.

After my roommate's departure I was struck by a profound sense of loneliness. I spent lots of time looking out the room's sole window, lowering its shade to different heights. When it was almost all the way down, the only thing visible was a strip of dry grass and gravel. I could watch the occasional leaf appear out of nowhere and scuttle across this scene, something arachnoid in its movements. When the shade was up, I could see a trapezoid of sky crisscrossed by sparrows and contrails, watch them cut and mend across the blue.

At night I lay in bed and it was like I was being pressed into the earth, flattened against its skin. I counted my breath in tens, pushing my lungs up toward the dark and then letting them fall back into my ribs. When I finally drifted off to sleep, I dreamed that I was boarding a train at the center of the planet, its tracks radiating outward into space. The journey began in the tunnel-dark of the core, silent but for the click of the tracks. After a while, the train emerged into the hot light of the mantle, and we chugged through it for hours, a monotony of rock and flame—all of it melting and crystallizing and re-melting—like we were passing through the guts of a lava lamp. Days seemed to go by this way, with nothing to see, barely a hint that we were even moving. In the dream I watched myself fall asleep in my berth, my head smudging against the window, my hands falling limp in my lap.

And in the middle of this ill-timed nap it happened: the sudden appearance of tree roots and basements, the eruption into sunlight, the forests and the waves and the billions of gangly creatures scuttling down streets and over seabeds, beating their arcs through the air. In a flash I watched the train shoot past the golden stupa, past the pines, past a sleeping figure lumped beneath a quilt in an unlit room. Then, in a few moments, a long blink, it was all behind us. I watched my dream-self wake up much later—again into a darkness, this one emptier, airier than the core—and realized with a pang that he had missed it all, that he was unaware there'd been anything else to see besides scorching heat or crushing cold; or that, even if he had caught a glimpse, he'd soon be compelled to dismiss the thin pellicle as something fleeting and forgettable, a strange and tiny exception to the suffocating default out his window, a night into which the train would plummet for years, maybe forever.

When I actually woke up it was dark in my room. The wind was at the windows and the sun was still hours from the horizon. For a moment I felt wildly exposed, lying there on the exact surface of the planet, suspended precariously in the critical zone between abysses, and it struck me that I'd only ever have a tree's worth of space—roots to crown, give or take—in which to raise high the roof beam and lower down the coffin and conduct all the affairs that might reasonably be said to comprise a life. The thought frightened me, and I spent the rest of the night taking shallow breaths under my quilt, trying not to do anything that might disturb the delicate envelope between ground and sky into which I had been sealed.

———

DURING THE RETREAT, our most important human interactions came in the evenings, when we would all gather in the meditation hall to watch a recorded video discourse delivered by a renowned practitioner of meditation, an avuncular Burmese man named Goenka. The man had been dead for half a decade, but in the videos he had heavy cheeks and neatly combed hair, and often emphasized his points by nodding his head from side to side or lifting a hand palm down in front of his chest.

I realized after a day or two that his lectures were reminding me of Latour's. Goenka's problem was less ecologic than psychic, but the goals seemed analogous. How, he was asking, could we learn to live with change? How could we avoid being buffeted by the constant swings between pain and pleasure that characterized our relationship to the world outside ourselves? To begin with, he explained, we had to recognize that there was no thing called our "self" that we could step outside of and point back to. Even our bodies were less things than processes, congeries of atoms constantly arising and passing. In order to grasp this, he said, we had to pay close attention to the sensations on our skin, the reactions of our bodies to their own activity. Every night he would give us the merest additional instruction to inform our next ten hours of meditation. After focusing on the breath in our nostrils, we were told to widen our attention to the area beneath our noses, then to each of our body parts in turn. I tried to take inventory of all the fleeting feelings on the surface of my body, the subtle shivers and pulses. The more I noticed, the more I seemed to notice, like my nervous system was a rough sketch whose detail was slowly getting filled in.

In the hours I spent doing this, I had time to draw out elaborate parallels. I thought about how I was attending to my body in the same way Latour would have us attend to the earth: not seeking any objective remove, but just sitting inside it, observing. I thought about how both of these processes require humility, that as with the critical zone, when you turn your attention to the skin you realize that you've been more or less completely oblivious to what's been going on literally under your nose. That you've been living inside something that you understand much less than you thought you did, whose apparent stasis obscured a disquieting flux. For Goenka, as for Latour, the prescription for this disquiet was simply further attention. Whether it was the surface of the body or the surface of the planet, their propositions were the same: the finer our awareness of the thing we were inside, the more resilient we'd be to its changes.

But though I tried to just inhabit this awareness, my attention kept wandering outward to examine its rationale, to try to read it from above. I often got so carried away in these trains of thought that I nearly opened my eyes, and it would take me several minutes to realize I'd lost track of my breath completely, let my thoughts on meditation replace the thing itself. The feeling of frustration this produced was similar to the one I'd felt watching Latour. Like I was being proffered a hand I couldn't quite grasp. Like I was missing the answer I knew was right in front of me. Goenka, for his part, seemed to anticipate this frustration. At the end of every lecture, he was careful to stipulate that the truths he'd been discussing could not be communicated all at once; they had to be stepped into little by little. It was for this reason, he told us, that meditation was referred to often as a practice, but never as a theory.

On the last day of the retreat I walked out to the stream to find that the wind had blown the plastic chair into the current. Its back was lodged against a rock and its four white legs stuck rigidly out of the water. The weather was fitful and the usually clear stream was congested with leaf litter and fallen branches, which piled against the upturned seat where I would have otherwise been sitting, watching them float by.

During his final discourse that night, Goenka touched briefly on the topic of death, ostensibly the greatest change of all. There is a famous painting of the Buddha's funeral, he told us, where the Buddha's body lies in state under a tree, surrounded by throngs of admirers and apostles. From out of this crowd, it was easy to pick out the people who had achieved enlightenment. "How?" he asked us. They were the only ones who weren't crying. After all, what was there to cry about? The Buddha's body had arisen and passed on, just as it always had, as everything always had. Something new would arise in its place. There was absolutely nothing to be sad about.

Latour, too, ended on a high note: "We should not be pessimistic about the situation," he said, stepping out of the light of his projections. "Of course it's tragic, it's a tragic situation." But it could also resemble a kind of Renaissance. Now is the moment, he said, for us to finally become citizens of the critical zone, to practice the kinds of attention it invites, to "discover a new way of being in the old land."

There is exciting possibility here, I think, though I have to remind myself that Latour's Renaissance will be different from Bruegel's. Less snow, surely, but something else, too. Because back then it was possible to paint the scene from afar, the whole village, the whole pond. Back then everyone was able to skate on the ice—the surface was that thick.

———

I WAS HOPING all this would be enough for us. That I could just find the Now, or the skin, or the vortex, and hand it to you here on the surface of the parchment, inscribed in a row of ink stamps. This was the whole purpose of my retreats: to find an answer, a means to break out of the cycle, so that at some point I'd be able to turn to you and say, simply, here it is. Here is how you live with the Problem, here is how you keep time. There is absolutely nothing to be sad about and/or I think you should still have hope.

But in the end, I've found there is something non-translatable about these sorts of transcendence, something that resists easy passage from one person to another. You cannot inherit Goenka's equanimity, or Latour's curiosity, or Tolle's presence, or Kirsten Dunst's stoicism simply by having them explained to you. It's the same caveat Wittgenstein gives in the first sentence of his *Tractatus*: "This book will perhaps only be understood by those who have themselves thought the thoughts that are expressed in it." Likewise, perhaps, with the Problem: that it will only be understood by those who have themselves felt the feelings it evokes— feelings that cannot be acquired from anyone else, but must be cultivated separately, in the privacy of each gut.

Maybe this is part of what Nelson means when she talks about the difficulty of achieving relation. That even if an answer is "right"—even if that rightness is somehow construed to be universal—it will still never be easily conveyed in a medium as naive as a letter.

Yet here I am anyway, searching for a mode of conveyance. I'm trying to relinquish the fantasy that I can simply confer a solution in prose, that this letter could ever really protect you. But still I seek relation—doggedly, foolishly—not despite but *because* of the fact that it's getting harder, that every year there are more pangs of doubt and tons of carbon and inches of rainfall coming between my generation and yours and it's like I can feel them flooding the wires, scrambling our lines of communication.

In the face of these facts I've grown stubborn. I don't want to stop calling across the widening divide, don't necessarily think that we should. Yes, it is muddled and reckless and perhaps prone to failure. But as the Problem widens the temporal gap—as it sorts you and me into separate geologic eras—I would still rather misfire in search of communion than just presuppose its futility.

Here is my best prediction: When the storms hit, anything that hasn't been battened down will be blown away. Everything empty will be filled and everything fragile broken. If your head is turned the other way, then at some point you'll be blindsided. And if the muscles in your heart have gone slack, then perhaps they'll be torn in the wind.

So if I cannot write you to safety, let me at least be clear about the limits of this letter. It will not put out any fires. It will not float or fly or shield you from the rain. It will not offer pithy consolations you can whisper back to yourself in your darkest hour.

There are those who insist they've already found the answer to the Problem; that their words have no limits, and they can therefore disengage. This surety takes many forms, all of them hollow. On the right they simply deny the facts: the answer to the Problem is that it is not happening. In some corners of the left, the response can be equally cynical: the answer to the Problem is that we're already done for. But though the temptation remains (for what could be more soothing than manufactured certainty?), I am not going to offer you this sort of easy, summative stance. "We're fine" and "we're fucked" are not answers, they're expressions of fear, walls we put up to avoid having to look at the Problem ourselves. And this is the point: that eventually you will have to look at it yourself, develop your own relationship to it. Eventually you will have to feel whatever feelings it evokes. It will be hard work, which is to say heart work— a process that must be experienced, that cannot simply be thought through or written out. But know that in this I am with you. That I have not yet felt the answer. That this is me, searching.

Third Movement

The pastor is standing with his back to the wall, reciting a eulogy. The weather is cloudy so the wall casts no shadow, turns no colors in the sun. It just looms there, impassive, rising well above the crowd.

This is the exact spot where the storm water broke through, though you could never tell by looking at it. The flood wall is inscrutable, each length identical to any other, a concrete stripe across the neighborhood. It is far taller than any of the shotgun houses that still remain in the lower Ninth Ward, so looking west down any street all you see is a gray barrier excising the view, like some sort of glitch in the sky.

What's surprising is the simplicity—you can just walk right up to it, press your palm against it, and lean. No signs, no fencing, no murals, just a run of stubby grass along its base. There are people who live right across the street in tiny rebuilt houses with low-slung porches, where every morning, if they want to, they can sit on their deckchairs and stare into its face. They cannot see beyond it, can't see the water at all. They can only look at its surface, that dull slab, and trust it to keep the bay at bay.

The pastor bows his head and begins reading out the names of the people who were killed by the storm. The list takes several minutes to recite. It's been exactly a decade since Hurricane Katrina made landfall in New Orleans, but around us the neighborhood is still mostly vacant, empty lots sprouting weeds, empty windows boarded in plywood. In some places there are old trees that survived the flood, though some of these are dead, standing only because they never fell. A mugginess sits over all of it, the hum of insects. The heat just stands there in the grass. Between the scattered buildings there are large gaps where neighbors used to be.

But there are also people who've returned, an act of almost unfathomable courage and resilience. Their rebuilt houses are little beacons in the overgrowth, all trim lawns and fresh paint, defiantly normal.

Walking to the memorial service I stop by one such building, a corner store advertising ice-cold Coke. Next to it is the gutted shell of a house, the words "no trespassing" sprayed in red paint on the bricks. The proprietor of the store motions me in with a wave, and I step through the strips of cloudy plastic hanging in the doorway. The inside is shabby but well stocked, the checkered linoleum peeling up in places. Only commercial business in the lower Ninth, the proprietor tells me proudly from behind the cash register. Featured on *Good Morning America*, CNN, and in the *Times-Picayune*. He points at a row of Xeroxed articles tacked and framed on the wall. The photos show him standing and smiling with his whole family on the sidewalk outside their store. The condemned building next door is not in the frame.

After the hurricane we just scraped the money together and

then opened up again, he says. People have to have somewhere to get food.

His wife comes out and offers me an electric blue shaved ice for one dollar. It's sweltering outside and I accept it gratefully, squeezing the paper cup straight into my mouth. She laughs and tells me to look in the little mirror they've got mounted by one of the coolers. And there I am, framed against the soda cans, teeth and tongue stained blue as dye. Happy tenth anniversary, she says.

This confused me at the time, though less so now. What better thing to celebrate than the ordinary, its unbelievable persistence?

It is 2015, two years after college and one year before I'll join NY Renews. I've traveled to New Orleans with a delegation of people who were displaced three years ago, when Hurricane Sandy hit New York and New Jersey. Though it took fewer lives than Katrina, Sandy's ongoing aftermath recalls what happened here in Louisiana: shell shock strangled in red tape, a slow sorting of debris. The Problem creates weird resonances like this, erecting new solidarities out of nothing, patching people together.

I've come to New Orleans as an organizer, with funding from the Sierra Club, for whom I still work. Since the success of the big march in New York City, the club has brought me on to run campaigns fighting coal plants and promoting offshore wind power. Though for this particular assignment, the work looks quite different. Mostly it consists of booking hotel rooms and arranging transport, whatever is needed to support the Sandy delegation. I arrange a few meetings, too, story circles with survivors of Katrina. We hold them in church basements and public parks in the days preceding the anniversary. There is a gentleness to

these meetings, an empathy borne of shared trauma. People nod
their heads and pat one another's shoulders. Strangers choke up
and then find ways to keep talking. There's an eagerness on both
sides to share lessons learned, though this is mainly another
form of catharsis. Most of the survivors have already had to learn
the basics: how to remove black mold, how to petition an insurance company, how to decide which of their belongings to keep
and to scrap.

On the day of the memorial service, the Sandy delegation stands
mixed in with everyone else, listening to the pastor list the names
of the dead. Most of those gathered are Black, as were most of
the victims of Katrina. Redlined over decades into the lowest-
lying sections of the city, New Orleans' Black communities bore
the brunt of the hurricane. When the water topped the levees,
their houses were the first to be flattened, their lives the most
likely to be erased. (This is a cardinal rule of the Problem: that
no matter where it hits or how, it will magnify the inequities on
the ground.)

The crowd of survivors is thousands strong, and everyone
has their heads bent in mourning, either down toward the ground
or up toward the sky, as if trying to catch a glimpse of the departed.
Staring down at my feet, I have a sudden flashback to childhood, a visit with your grandparents to the Wailing Wall. Another
monolith at whose feet people bowed, a remnant of tragedy transformed into a cenotaph. I remember how the black-hat men would
approach the wall and touch it with reverence, as if the pitted
sandstone held something of their fate, as if rippling behind it
were the outlines of the world to come. They would press notes
into its cracks, wedging them between the twigs of caper bush,
willing their scribbled prayers to come true.

Staring at the cement of the flood wall, it too seems premonitory, though of what I can't tell or don't want to. It feels now—looming over us in the heat—like both a protection and a confinement, though perhaps you can't have one without the other. Some people in the crowd do walk up and touch it, or lean their backs against it, listening to the pastor. But there are no cracks in it anywhere, no room for notes, just a blank face that everyone hopes will hold.

When the pastor concludes, a change comes over the crowd. People start milling about, stretching themselves into the beginnings of a march. Several brass bands appear and move toward the front, hefting tubas and trombones over everybody's heads. You can see them through the crowd, catching the sun, a glistening thicket of horns. And then without warning they start in, and everyone is moving—forward, side to side, up and down on the balls of their feet. It is less a march than a dance, flushed down the street by trumpet and drum. As the procession passes through, it's like the neighborhood is reanimating. People open their windows and step out on their porches. They overflow into abandoned lots, and dance on top of parked cars. There are people in feathered headdresses and old varsity jackets and cut-sleeve denim vests, everyone swinging their bodies around, sweating from the forehead.

At the story circle we'd organized earlier that week, a Katrina survivor had explained that this was another type of mourning. What was planned was not a parade, she'd told me, but a second line, a jazz funeral without a casket. The tradition was centuries old, the hybrid of Western military processions and the West African dances brought to Louisiana by enslaved people. In this case, ten years after the storm, it would commemorate all those who had lost a home or a life.

So even in a long history of second lines, the one I'm stand-ing in now is one of the largest in New Orleans history. As the music picks up speed, the march leaves the Ninth Ward and makes its way downtown, passing over highways and down side streets lined with dollar stores. The music is so loud and infec-tious it's like it might actually raise the storm-dead, summon them from the water to dance. By midday, the sun has burned off the damp and there are no longer any tears; the cheeks of the marchers have all dried up and cracked into grins.

There is wisdom in this alchemy of grief into euphoria. The way it sees death and raises it life. The way, where there is no happy ending, it creates one. Together the marchers dance through the city for what feels like hours, dance until the sweat stains our shirts and the brass bands are gasping for breath, and then we keep dancing anyway, in celebration of our bodies, in defiance of them.

When we finally reach the end of the march, we are standing in a large grass field between highway on-ramps. There is a stage at one end and various people have started to address the crowd. One of them is a director at the Sierra Club, a white man in a green T-shirt. He begins referring to the Problem by name, try-ing to frame the whole day in terms of it, but no one is really listening. People fan themselves and pour water over their heads, retreating to the shade beneath the overpasses. At the back of the crowd, I buy a roast turkey leg from a man with a barbecue and prop myself against a chain-link fence to eat it. I notice I feel vaguely embarrassed by the man on stage, like he is missing something, striking the wrong note.

And I didn't know this then, but it's something I realize now: that sometimes the name of the Problem can erase the thing it-self. That there are moments when the Problem should not be contained in a word or phrase, when in fact language backfires, and the thing it is trying to animate winds up flattened, stationary, remote. I've come to understand this only after years of making the same mistake, trying to use the name of the Problem to evoke the experience of it, like trying to capture the whole of a book by invoking its title.

The anthropologist Kathleen Stewart talks about "public feelings," affects that "begin and end in broad circulation." Try to localize them in a single person, a single word, and they're lost to you. "They work not through 'meanings' per se, but rather in the way that they pick up density and texture as they move through bodies, dreams, dramas." This is, I've begun to think, how the Problem works, what it *is*. The Problem has never been just a single thing, to be pinned down and abstracted under a title—as if any one name could contain the contours of a hyperobject. There is just too much it causes, too much that is its cause. It implicates entire political and economic systems, widespread ontological and epistemological norms, the fine-grained frictions of daily life itself. Which is why, I suspect, the full feeling of it can only ever be public, located in the gaps and sinews between things, perceptible only in constellation. Not something any of us can "have," but a truth wrought together through hunches and glancing blows.

That day the Problem could not be found on the stage, or in the microphone of the man who employed me, or in any of the dutiful words filing out of his mouth. It was elsewhere, everywhere. In tensed shoulders, patched roofs, the pucker of lips to a trumpet. In the memory of paperwork and food queues and foldout cots. How there was never an evacuation plan in place for people without cars, and how when the storm hit it was the end of the month and everyone on welfare was strapped for cash, and how the morning after, as the flood wall cracked and finally gave, the president was on TV declaring that New Orleans had "dodged a bullet."

The problem with my director's speech was that the crowd already knew about the Problem, understood it on a level deeper than its name could ever penetrate. And when he finally finished, people emerged from the shade to dance again, trampling the grass into some kind of prayer, willing it in their bodies to come true.

For this reason, as you've probably noticed, I've chosen not to give the Problem one of its usual names. As Stewart puts it: "This is not to say that the forces these systems try to name are not real and literally pressing. On the contrary, I am trying to bring them into view as a scene of immanent force, rather than leave them looking like dead effects imposed on an innocent world." Maybe this is a better way of expressing what it is we were trying to do with our marches: to bring the Problem into view as a scene of immanent force. To literally embody it, even if just for a couple hours.

In Stewart's schema, the public's feeling about the Problem mirrors the carbon it propagates: diffuse and ubiquitous, completely invisible until it's set a hillside aflame (or thousands of people to dancing). Though perhaps "feeling" isn't the right word here—maybe "mood" is more accurate, something infectious, like laughter or disease.

Unlike a feeling, a mood can't be captured from a single vantage. In this regard I'm like the proverbial blind man, describing only the portion of the elephant that he himself can feel. Except instead of a trunk and a tail, the hyperobject has infinite limbs, unlimited manifestations. I doubt we'll ever render it fully or capture its bulk in the net of a name. But there is comfort at least in knowing that there are others out there as lost as you are, fumbling across the surface of something too immense to hold. That's one premise of this letter. That there is value in sharing what it feels like from wherever it is you're standing.

———

ON OUR FINAL DAY in New Orleans, the Sandy delegation is invited to take part in Sunday services at a rebuilt Baptist church. The church is more than an hour outside of the city, southeast along the bird-foot delta that steps delicately into the sea. The land here is just sediment, washed out of the mouth of the Mississippi and deposited in a loose fan that, even at its highest point, rises only a few feet above the water.

We drive down the peninsula in a rented minivan, past ranch homes lofted on thirty-foot pylons and billboards advertising sport fishing expeditions. We drive past a new football field, incidentally the unit of measurement people use to describe coastal erosion around here. One football field lost every hour, apparently, though this one's been shored up with a low earthen berm. The process is driven as much by sea level rise as by the offshore extraction of oil, which causes the land itself to subside—a vicious, self-defeating loop, the autophagy of the state's economy (though the oil companies make off handsomely).

Katrina is still here, too—not as wreckage but as absence. The whole area looks like it's been wiped clean: of trees, of towns, of most structures. Even on a sunny day I feel exposed somehow, like we're driving across the surface of a raft. The ground is too flat, the ocean too close. We round a bend and there it is, only a few feet from the wheels, almost exactly level with the land. It is frankly intimidating, like something that could eat you but happens not to be hungry.

After a couple of hours, we arrive at the church. It is small, stilt-less, new. The walls are covered in prefab white siding, and a modest steeple rises from the center of the roof. The pastor greets us outside, shaking each of our hands in turn. Inside, the building is also white, and the congregation is dressed in immaculate church clothes, little brooches, and polished shoes. We take our seat in the pews, holding our prayer books open in our laps until it is time to stand up and sing.

The pastor's sermon that day does not mention the Problem by name. Instead he names the storms, Katrina and Sandy, invoking them like something out of the Bible, like two angels of death summoned to lay waste to the delta. But by the grace of God, he says, we have survived, and our church has been resurrected. And now we sit here once again to deliver praise unto Him; He who will greet us when it is our time to ascend.

As he speaks I picture the flood maps, how the place where we now sit will be consumed in under a century, erased by the body of water known as the gulf—and how indeed this will represent the expansion of another gulf: between here and there, between past and present, the two no longer separated by a mere sequence of events but by a literal change of phase, the difference between earth and ocean. And how everyone in the congregation—or their children, or their children's children—will ascend north, and how the buildings that used to be their homes will fade beneath the surface of the gulf, and how passing over them in a boat decades later it is possible that even the peaks of their roofs will be invisible under water that's been clouded by successive spills of the oil that banished them in the first place. And suddenly everything—the land, the church, the sermon—feels charged with this disappearance, like there's another world welling up behind this one, and I grip the edge of the pew as if it might float away and then everyone intones amen.

Afterward, the pastor invites all the Sandy survivors to the front of the dais. He places his hands on their heads one by one, and in a voice that is so warm and gentle it is practically a whisper, he says, "Grace be upon you."

When the service ends, people step outside into the sun, striking up little pockets of conversation around the parking lot. They loosen their ties and take off their heels. They exchange hugs and hold them for a beat. Babies get hefted from strollers, children weave in and out of legs. On a screen somewhere far away there is a pixelated blue eating into the edges of this scene, but for now the impermanence feels like a kind of ecstasy, and the morning washes out into a long afternoon. When we finally leave—waving goodbye and sliding shut the door of the van— there is a sense that the grace is right there; that, as with so much else, it is already upon us.

Object

When I was seven we caught a salamander in the stream near our house. The stream crept unnoticed through the verges of our neighborhood, descending inexorably toward the polluted river. To find it, you had to walk to the end of our block, where a yellow sign with a double-headed arrow indicated each of the two directions you could go. If you ignored the sign, which on that day we did, and ducked under the fence behind it, you arrived at a trickle of a creek that disappeared quickly into the mouth of a corrugated-iron drainpipe.

My sister and I were fixated on the idea that something lived in this creek and begged our babysitter to investigate with us. We found a small net and bucket in the basement, and climbed down through the bushes, out of sight of our house. The babysitter dipped the net in and out of the water, while my sister and I crouched on the banks, clenching our fists in anticipation. Our legs had started to cramp and prickle by the time the babysitter pulled up, to his surprise and ours, a small brown salamander. It wriggled vigorously in the net, and in our excitement we imitated its panicked, gleeful little dance.

We went about collecting water and stones from the creek, and at home we placed these and the salamander in a large Tupperware with a lid we'd punched full of air holes. The whole "habitat" was displayed proudly on the front table, next to the Shabbos candles. Then the salamander disappeared. Looking through the sides of the Tupperware, we could detect no movement—the salamander could have been any one of the tongues of mud pressed between the stones. This went on for several days: my sister and I checking and rechecking the Tupperware, waiting for some event to confirm the existence of the thing we knew we had caught. The salamander waited us out, denying us the satisfaction, venturing not even a flick of its tail. I asked my mother why the salamander wouldn't move, and she told me, quite reasonably I thought, that since we'd put it next to the Shabbos candles the salamander was napping. "It thinks every day is a day of rest," she said. When Friday night did come, the candle flames cast slithering, amphibious shadows all over the stagnant tub.

At last, we relented and brought the Tupperware back to the creek. I held the container at an angle in the water and my sister peeled off the lid. And though we emptied all of its contents, we never saw the salamander leave. Just a cloud of mud and then the creek as it was.

As I grew older, I came to understand that this was not a fluke. Disappear was what animals *did*. My relationship to them—to the extent that it existed at all—was strung together from glimpses, the sudden seconds before they darted into a bush or scurried to the back of an enclosure.

This was hard for me to take, especially as a child. Weaned as I was on Disney—where animals could be your sidekick or your nemesis, your best friend or your comic relief—I was surprised to learn that the denizens of the New Jersey woods wanted nothing to do with me. On those first hikes, I remember feeling like a pariah, chipmunks and deer sprinting through the undergrowth to get away from me. They would always freeze for a second first—bodies tensed, eyeballs swimming in their heads—before beating their abrupt retreat, leaving me squinting after them into the trees. The only way to prevent it, I found, was to stand still and make almost no noise—to pretend, in essence, that I wasn't there at all.

By the time I entered middle school, this was no longer cause for much surprise. By then, I had undergone the predictable loss of innocence that I assume befell many of the children who came of age in the waning decades of the twentieth century. The realization, specifically, that all the fantastic animals we'd learned about in science class, the ones we'd penciled in in our coloring books—the pandas and jaguars, the orangutans and the elephants—were being wiped off the face of the planet. That their rain forests were burning, their savannahs were being cropped, and that we were the agents of this extermination.

In this light, it made abundant sense to me that most animals—even the more prosaic species of New Jersey—would do everything in their power to avoid us. I began to feel rueful every time I encountered so much as a squirrel or a sparrow—as if their scurrying retreats were more than just an instinctual response; as if it was all in fact a very deliberate snubbing, an animal kingdom–wide pact to shun the humans in protest of what we'd done and in fear of what we might still do.

Later, I learned the precise scope of our obliteration: humanity—or, more specifically, that blinkered strain of economic thought that had transformed all ecosystems into "resources," to be harvested or paved over—had triggered the sixth mass extinction since the advent of biological life. This was a deeper kind of disappearance. Die-off rates were a thousand times above normal, and dozens of species—entire species!—were being snuffed out every day, punching holes in my picture of the planet. I felt aware every time I got into bed that I'd lost an irretrievable opportunity to witness a whole host of creatures that I hadn't even known existed, and that the same thing would happen tomorrow and the next day and so on into infinity.

This made me feel very sad, though at the same time I took it in stride. The avenues available to me for learning about the nonhuman world—outlets like National Geographic and Animal Planet—were always focusing on the "last remaining X" or an "extremely rare Y." Though their coverage was usually meant to promote conservation, the message I took away was that I should avoid getting too excited about the Javan rhinoceros or the silky sifaka, as I'd likely never see one, and there might soon be none left to see.

Implicit in all this was an equation of scarcity, which I internalized early on: the more remarkable an animal, the likelier it was to vanish. This was unsurprising to me, probably because I'd already osmosed a similar lesson growing up under late capitalism: that scarcity produced value and vice versa. So although it was sad, it made sense to me that Javan rhinoceroses were disappearing. They were too good to last, and this seemed less a consequence of the Problem than a guiding principle of the world.

Now I look at the sheer number of extinctions and they appear to portend the end of the world, though this isn't how the Problem works, exactly. The Problem doesn't operate at the level of the "whole world," which anyway can't be reduced to a singular thing, a story to be concluded climactically and all at once. It finds its traction instead in the smaller worlds, the subworlds: a system of brackish creeks draped in mangrove; a narrow band of dwarf spruce huddled just below the tree line; the heady two weeks between the birth of a bee and the wilt of a small red flower that only it can pollinate. Worlds that often pass beneath our notice, but which have for millennia made up the absolute outer boundaries of subjective experience for the many other species with whom we share our planet. These are the worlds that the Problem is ending, blinking out one by one in messy, staggered succession.

Like most people, I've grown accustomed to these blinkings, and am now almost completely numbed by statistics on extinction and biodiversity loss. They shocked me as a child, but I can no longer fathom their accrual, cannot make them mean anything. Instead, what I'll sometimes try to do is inhabit a single example, an apocalypse in miniature.

Recently, for instance, the snow that once blanketed the Alps (you can see it in the upper-right-hand corner of Bruegel's painting, softening the peaks beyond the town) has begun to melt drastically, even in winter. I've read that there is a species of hare living in these mountains that has evolved to turn its coat white in the winter as a form of camouflage. When Bruegel sat down in 1565 to paint *Hunters in the Snow*, dozens of these hares might have been nosing their way across the scene, entirely invisible to him.

I imagine them traversing an environment of simple slopes and planes, padding across fragile crusts of snow, squeezing down blind tunnels lit by a cold sun filtered through ice. Theirs is a glittering, muffled world, where every sound is both faint and pronounced, like a single stroke on a blank canvas. All movement here is dangerous, so they try to stay frozen—an immobility belied only by breath, discernible in the slight lift and sigh of the fur on their haunches. What food there is they exhume from beneath the snowbanks, small portions of freezing grass shattered down the throat. Centuries pass this way: silent, furtive, white on white.

Then the whole world, the high white world that lent them its color, begins to break into pieces. The snow comes but it doesn't last, melting first on the meadows, then receding to the divots and trenches of the mountain. What had once been a single expanse of camouflage is suddenly a chain of disconnected islands, each of them steadily shrinking, and against the bare ground the hares stand out like targets. It's as if the mountain, once their safeguard, has betrayed them, disclosing the secret of their vulnerability. Dressed for a world that no longer exists, the hares dart between the remaining drifts, paws crunching on gravel and splashing through puddles of snowmelt. Everything is muddy and coarse now, entirely too loud.

In this photonegative country, the wolves devour the hares, snatching them from the mud in a frenzy, too full to keep eating. Perhaps in the shocking ease of the hunt, even the predators sense that something is off, that the feast can't last—that as the hares begin to disappear, so too will they, for the simple reason that the same snow protecting the hares from their hunters has protected the hunters from their own appetites. In the wake of this localized apocalypse, the hare's world would grow truly unrecognizable: warm, empty, devoid of both safety and danger. Though from Bruegel's vantage it would look more or less the same—scenic and remote, maybe just a shade browner.

There is a near endless supply of these examples. In the cloud forests of Costa Rica, the heat pushes a certain species of tree frog up the slopes of the mountains. They climb higher and higher in search of cooler temperatures, but the elevation runs out and the temperatures keep rising. In my head I picture these frogs poised on the highest branches of the tallest trees, stranded on their dwindling peaks—how their skin would slowly dry to a husk, pulling their mouths open and their eyes wide, ossifying them into gargoyles.

Or conversely, I picture the pine forests of Vermont, where warming winters are unveiling new territory, allowing ticks to push northward. (Ticks being, along with disease-bearing mosquitoes, one of the few species I've ever read about *benefiting* from the Problem. Which lends an eerie sense that there is something inherently pestilential about it; that it doesn't simply adhere to the cascading rules of physics, but is additionally propelled by something deliberate and retributive, like a biblical plague.) In increasingly large numbers, the ticks crawl through the tall grass, attaching themselves to moose and boring into their hides. I've read about a dead moose found with ninety thousand ticks on its body. I've read that some moose scratch so hard in an attempt to be rid of the itching that they rub parts of their skin off on the trees. Their numbers have shrunk dramatically in recent years, felled by infections and blood loss, so that their likenesses—still stamped into maple candies and emblazoned on postcards across Vermont—are increasingly tinged with nostalgia, as if every souvenir were a little memorial.

Even as I picture them, though, I am wary of the way these stories are deployed, how they are passed around like a form of currency, a tender of loss. I know there is a long history of environmentalists using charismatic animals-on-the-brink to pluck the heartstrings of their mostly white, middle-class donor bases, and that these appeals have often subsumed or replaced the voices of the millions of poor *people* at risk from the Problem. But I also can't dismiss them entirely: to me these stories have a meaning deeper than fundraising, deeper even than sadness. Like that scene in a movie where all the pigeons fly away before an earthquake, or where the rats dash down the drainpipes before a big flood, I think these are fundamentally stories about fear, about being left alone in a dangerous place. Though they are often packaged safely into bite-size tragedies, taken as a whole they look more like a jittery, collectivized unease, a vague inkling that we're being abandoned on an increasingly empty planet.

In a way, this is good: we are long past the point where we should be putting starving polar bears on wall calendars, pitying them like their fate has no relation to ours. The rhetoric of trying to "protect" animals serves only to imply their expendability, casting them as fragile, aestheticized treasures so we can avoid seeing them for the bellwethers they are—less worthy of pity than of panic. Now when I read about extinction, this is something I try to hold on to: that in every regrettable loss there is a seed of paranoia; that at some point their disappearance may start to feel like a desertion.

———

IN COLLEGE I BECAME FIXATED ON ISLANDS, the most obvious of endangered subworlds. Tarawa, Majuro, Kwajalein—all the tiny motes of land that were disappearing into the sea. Sitting in my carrel in the library, I would scroll through aerial photos of far-off atolls, thin donuts of land balanced precariously on the skin of the ocean, as if held up by surface tension alone. I'd watch videos of their political leaders, who managed somehow to contain their desperation and give poised, impassioned speeches at each successive UN conference, urging the Global North to contain the Problem before it wiped their countries off the map. They mustered a candor that no one else could, brought the hyperobject into new and terrifying relief. I gravitated in particular to the words of Tony deBrum, then minister-in-assistance to the president of the Marshall Islands, who was always heroically frank: "We must all declare our sincere and heartfelt commitment to take action, every one of us—government and citizen alike—to win this war. This is a war for nothing less than the future of humanity. It will take every one of us." The Marshallese poet Kathy Jetnil-Kijiner put it more personally: "Tell them what it's like to see the ocean level with the land / Tell them we are afraid / Tell them we don't know all of the politics and the science but we see what is in our backyard / Tell them that some of us are old fishermen who believe that God made us a promise / Tell them some of us are a little bit more skeptical of God / But most importantly tell them we don't want to leave / We've never wanted to leave."

There was a profound sadness in listening to these speeches from my library carrel halfway across the world, though this sadness was often accompanied by a sense of foreboding. To me the islands felt premonitory and microcosmic: entire subjective realities—bounded, unique, and dense with meaning; home to hundreds of generations of unreplicable human experience—faced with the simple and unfathomable prospect of elimination.

They seemed like windows into an ultimatum that the Problem might eventually deliver to the rest of the world, in one form or another, over the course of the next few centuries.

All of this was by necessity an armchair anxiety, and since I had no means of visiting a low-lying island, I assuaged my feelings by collecting as much information as possible, as if personal knowledge of these places would help prevent them from sinking into oblivion. In my sophomore year, I signed up for a research project with a professor who wanted to figure out how many species were at risk of extinction due to rising sea levels. Our methodology was to compile a list of species that were endemic to low-lying islands, and then find the exact elevations of those islands to determine which would be inundated under various sea level–rise projections.

We had no trouble finding dozens of species. There was a delicate-looking daisy that lived only on Starbuck Island; a sub-species of coati that lived on a cay near Belize; even, on one island, a unique shrewlike rodent. But current information on exact elevations was harder to come by. After weeks of more or less fruitless online research, I ended up spending a day in the basement of the Harvard Rare Maps library, poring over old survey maps wrested from the Japanese during World War II.

The titles and legends of the maps were all in kanji, and even the islands themselves looked like a form of abstract script, variously accented *I*'s and *O*'s written north to south on a grid-lined ocean. Often I couldn't tell which island I was looking at without comparing its particular coastline and archipelagic position to the images on Google maps. Most of the maps did, however, have topographical lines in five-meter increments, and I spent hours tracing a protectively gloved finger along these contours, finding tiny printed numbers, which I recorded in a spreadsheet.

When I finally left the library, I had collected elevation data for dozens of tiny islands I would never visit.

After bringing these figures back to the professor, I didn't hear from him for months. Eventually, once I'd all but forgotten about the project, he emailed me to come to his office. The data wasn't good enough, he told me, looking sheepish. Five-meter increments were too broad. The sea level–rise projections were all measured in singular feet, and we would need more granular elevation data in order to convincingly determine which islands would be submerged under each scenario. Apparently there was a government satellite that was slated to do a global elevation survey in the years to come, but there was no word on when its findings would be made public. Meaning that some of the islands might actually be gone by the time we found out just how little they'd once risen above the sea.

This scared me: the realization that the Problem could beat us to the punch like that; that it could advance so quickly that it might outpace our efforts to comprehend it, to even so much as count our losses. It felt like living through the Rapture, certain creatures being sucked up into the sky while the rest of us scrambled around here on Earth, trying desperately to make sense of the disappearances.

Years later, this feeling followed me to New York, where along with two friends I'd moved into a quiet apartment across from a laundromat. Twenty minutes north of us was the Bronx Zoo, which we'd heard was one of the best in the world. We decided to visit on a Wednesday afternoon, and though by then I'd already started the job with NY Renews, I made the decision (rare, guilt-ridden) to call in sick for a few hours. It was late fall and the weather was dingy, rain spitting out of a low sky. We took the train north, watching the borough sweep beneath us in a blur of brick and asphalt, knots of highway, a mud-choked culvert hemmed in by salvage yards. The zoo, like many of the parks in the city, had the feeling of having been carved out from the bustle of the city, like it was at pains to keep the surrounding neighborhood at bay. Inside we could still hear distant car horns, which mingled weirdly with the huffs and chitters coming from the animal enclosures. Otherwise, the zoo was relatively quiet that day, with few people around save for a weekday assortment of seniors, drifters, and nannies with kids, most with their hoods or umbrellas flipped up against the drizzle. The trees lining the paths were all shedding their leaves—small, yellow ovals that fluttered to the pavement and were plastered there by the rain.

Wandering around we came upon something called the Wild Asia Monorail, advertised as a sort of train safari. The setup was similar to the subway: we all took our places in the vinyl bucket seats and faced outward toward the scenery. Apart from us, the conductor was the only one in the car, a young woman wearing a visor, a headset, and a poncho several sizes too large. The train crawled to life and began snaking through the woods, about twenty feet off the ground. With no glass to cloud our view, and only the slightest hum coming off the rail, it felt almost as if we were sitting motionless while the landscape scrolled past us like a film reel. Occasionally, the top of an apartment block would appear through the foliage, breaching the Wild Asia facade, its fire escape piled with laundry or old bicycles.

We passed what looked like several small animal enclosures, but whatever was in them had taken refuge somewhere out of view. The conductor was narrating our journey with a litany of facts about the animals we weren't seeing, speaking into her headset with the scripted singsong of a flight attendant. "And up ahead we have the rare Przewalski's horse," she said, as we scrolled into an enclosure where a few droopy-looking ponies were huddled in a corner. "Known for their stocky build and short legs, these horses are almost extinct in the wild!" Something about her mechanical diphthongs—how cheerily impervious they were to the weather, the dearth of animals, the looming threat of extinction—made us cringe with laughter, and we had to pull our hoods around our faces so she wouldn't notice.

I remembered a point Emily liked to make about the status of animals, particularly zoo animals. How it was surreal that children were still given plush facsimiles of bears and lions to cuddle with in bed. How alongside the names of fruits and vegetables and common household items, we were still using words like "elephant" and "hippopotamus" to teach them the alphabet, as if these were relevant enough to be included in their very first batch of nouns. Except unlike with apples or basketballs, almost no children born today were ever going to encounter an elephant or a hippo, whose dwindling wild populations were increasingly being dwarfed by their thriving numbers in cribs and kindergarten classrooms.

Yet somehow the animals lived on anyway, signifying our sports teams and selling our breakfast cereals—vestigial symbols bereft of their original referents, floating loose now through our cultural imaginary. The word "bear," Emily liked to point out, originated from the proto-Germanic *bero*, meaning simply "the brown one." This was the euphemism employed by the Germanic tribes of northern Europe, who were so terrified of the animal that its true name was taboo, invoked seldom enough that it's been lost to history. What connection did this "bear" have to the symbol that five-year-olds now cradled in their arms before sleep, or watched dance to the Finger Family song on their iPads? What could the word even be said to mean?

After a few minutes, the monorail towed us past the horses—which hadn't looked back or even registered the presence of a large, empty train hovering over their corral—and then took a long curve through some trees. "Here you can see the majestic Malayan tiger," the conductor droned, as we pulled into another seemingly abandoned enclosure. "Check out its distinctive orange and black stripes!" Searching for the supposed stripes on the tiger that wasn't there, we really started to lose it, and the conductor shot us an embarrassed glance from under the brim of her visor. Then we all just sat there for a moment, the train paused on the tracks, the rain picking up speed. There was a single red stoplight visible through the trees; there were two hours left till the zoo closed; there were three hundred Malayan tigers left in the wild. There was no movement at all from the enclosure, save for the small yellow leaves, which fell down in gusts, catching on the barbs of the fence.

And then, right as the train was pulling out, I leaned over the edge and caught a glimpse. I could just make out the edge of its back, healthy-looking but thinner than I'd expected, shoulder blades visible through the pelt. It was slinking along directly beneath the monorail, precisely out of sight, denying us the satisfaction.

———

IN THE NEW JERSEY OF MY CHILDHOOD, the Rapture had already run its course. Many of the original species had left long before I was born. Instead there were streets and sidewalks, borders and zones. There were areas designated commercial and areas designated residential and small, scruffy parks set aside for recreation. The layout of the state—or at least our section of the state, a solid, suburban sash across the middle—was so monotonous that it seemed better represented by summation than by elaboration.

Picture a chain store—it could be fast food or an auto repair shop—swimming in a moat of asphalt. The building is all glass and rounded plastic, glossy but worn, like a child's toy left out in the sun. Behind the store, where fewer cars are parked, the lot ends in a chain-link fence. There's a dumpster in the picture, with one lid open, but no trash in it, just the fetid metal bottom, and the fence juts out around a transformer box, which has been painted an unassuming green and whose doors have no visible handle. At the base of the fence the weeds have been whacked and their severed stalks left there in drying clumps. Immediately behind the fence is a swale of coarse gravel, and behind this is the woods. Not a forest, but a thickening of trees, dense but sort of depthless, undergrown with brambles and shoots. All the trunks can be encircled with two hands, and the bushes are brownish and provisional. The word I think of is "scraggly," but this doesn't capture it exactly. It's that the woods feel both brand-new and already old, like they've aged prematurely. There are crushed cans and acorns on the ground, torn labels smudged into packed dirt. There is a small stream audible just out of sight. There are birds chirping, a mourning dove.

As you can see, it's not all bad here. In fact, it's kind of peaceful, in the way that all neglected places, freed from the expectations of human attention, seem to slacken into themselves, develop an atmosphere of their own. The New Jersey I remember was a patchwork of these spaces, always hemmed in, always hidden in back of something else. The marshy area between two subdivisions, the graffitied sound barrier behind the baseball field, the little-used picnic table behind a certain big-box home-goods store—all of it linked up into a vast interstitium, a here with no heart, only layered peripheries. To me these places felt sacred, possessed of some strange spirit that had emerged to fill the void of our neglect. Together they comprised a second growth wilderness: surrounded by and yet completely outside the slipstream of human concern, like so many islands in a river.

No one spoke about them, these little worlds emerging under our noses. Our attention was always directed outward, toward some distant place of astonishing beauty. In school we were made to read things like *Walden*, which I came to hate, not only for the tone of Thoreau's prose—which I found preachy and reproachful—but also because the whole thing felt irrelevant. Where was this cabin we were supposed to escape to? To what untrammeled glade could we realistically retreat? It felt like an insult, stoking this desire for pristine country in a world (Acela Corridor, the late aughts) where it was thoroughly inaccessible. Of course I'd read Krakauer's *Into the Wild*, of course I harbored fantasies of escaping to Alaska or Kamchatka or any one of those map points that retained its mythos as a redoubt of the receding wild. But at a certain point it began to feel futile and maddening— I didn't want any more stories about the wondrous world we'd missed out on. I wanted a book about the world I knew, the one I actually had to live in.

No such book was assigned, though now at least I have a better sense of what I was after. By then I was already feeling the need for a language of devastation and loss, a language that had loosened its grip on purity so it could begin to glorify what was left. What I wanted—badly, inchoately—was a kind of scraggly aesthetics: the ability to discern, amid the many endings, some trace of a beginning.

Outside of school I went looking for these traces, though they were rarely where you expected them to be. Once, I biked with a friend up the Watchung Ridge, a steep-sided crest that runs the length of north-central New Jersey. (The name derives from the original Lenni-Lenape *Wach Unks*—meaning "high hills"— though we pronounced it more like "watching.") Our hope was that from the summit we might witness a view that would make New Jersey beautiful, as if beauty were a pattern that would emerge only at a distance. But when we cranked ourselves to the top and looked out, the pattern below us was less beautiful than stifling: more parks and billboards, more highways and roads, with pulmonic little tracts of housing curling off in branches and these branches forming into larger clusters and these clusters spreading up the slopes of farther ridgelines, which receded west toward the horizon they obscured. What struck us more than anything was the repetition, like zooming out on a fractal—how, even from this vantage, nothing particularly new was revealed, just an endless reification of the world we already knew. Even the clouds had that tessellated effect, hundreds of identical white tufts plastered to the ceiling of the sky, fading out into perfect redundancy. The whole thing induced a feeling of claustrophobia, as if the vista were a trap that we'd sprung on ourselves.

As a consolation, we found our way to an Italian-ice place in a strip mall at the base of the ridge. We both ordered smalls and sat down at the picnic tables facing the parking lot. I'm still not sure how it started—maybe we were tired of straining our attention toward some distant place of astonishing beauty, maybe we were searching for something closer at hand—but at some point we began to notice the parking lot islands.

I don't know what else to call them, those manicured atolls of curb that guide traffic through a lot. One was shaped like an *L* with a single sapling at its joint, growing up from a bed of rounded stones. Another was long and narrow, with a low green hedge and a synthetic-looking mulch. There were several that were just gravel, lifeless beds in various amoebic shapes, with little pseudopods sticking out to delineate the different zones in which a car might park.

In jest at first and then a little bit in earnest, we started walking from one to the next, taking in their details. Our favorites were the most restrained: a rectangle of crabgrass with a stripe of dirt, an inscrutable hillock of loose stone. We took close-up photos with our phones, crouching down here and there in the parking lot, framing each island within its border of curb and each curb on its canvas of asphalt. It felt refreshing to be zooming in instead of out, to be looking closely at where we were instead of straining to find an elsewhere, some forest beyond the fractal where we were meant to find a cabin. Every island we stopped at looked like a tiny world, completely bounded and arbitrary, with its own structures and inner moods. And maybe this was all just a symptom of our metaphysical desperation, but the very fact of their banality—the fact that, though they'd been built by people, they now effortlessly evaded all attention—seemed to us to imbue them with the same quiet obscurity that Thoreau had presumably sought when he'd set off into the woods.

My friend joked that we should compile all our pictures into a coffee-table book, with full-page glossy photos of each island, maybe some inspirational captions. *Parking Lot Isles of Greater*

Watchung, we could call it. Nothing ever came of this, obviously, but I'm being serious when I say the book has stayed with me as a prized hypothetical, an aesthetic antidote to *Walden.* Because what frustrated me most about *Walden,* I realize now, was that it attempted to simply escape the Problem. Though no one knew it then, the Problem was of course already under way by the time of Thoreau's writing, and the things he described wanting to escape—the constant work, the dearth of time, the "lives of quiet desperation," not to mention, more tangibly, the rapid industrialization that had left him feeling familiar only with a "maimed and imperfect nature"—are recognizable today as early symptoms of its advance.

The point of our imagined book was that Thoreau's recourse wasn't available to us. As far as we could tell, there were no projected futures or protected glades where the Problem wasn't happening. Escape was impossible, in other words, and we no longer felt capable of maintaining the fantasy that it was. What was left, then, was to make our home in the world we had—or rather, to look for new worlds within it, unexpected nooks that we could unfold and make matter.

Stopping in front of one of the islands, I tried to imagine that I was a very small inhabitant, stranded there in an asphalt ocean. How each of the few objects that composed my world would take on a commensurate importance: an unusual duning of sand toward the southern edge, a splintered field of mulch on the eastern, not to mention the oblong shrub itself, which would, in my new ontology, assume an almost unspeakable significance. I imagined familiarizing myself completely with the island—every gradation of soil, each small bit of bark and trash—until its contours became like those inside my head, private and rote, wholly a part of me. I can't explain it more than this: I wanted to live there. To contract my horizon until I resided not in a world of redundancy, but in one of utter uniqueness—an island so buoyant with meaning that it would never, ever be submerged.

———

WHEN I WAS GROWING UP, we owned a coffee table book—an actual one—with sumptuous photos of the national parks. I was not immune to Thoreau's fantasy, and would spend hours on our living room sofa, poring over its pages. Here were the distant places of astonishing beauty, the islands of old growth that had been preserved amid the new, circumscribed by boundaries set down in law. In my head they played ready foil to the world I knew in New Jersey, and though a part of me felt suspicious of their allure, I was also desperate to visit.

When I finally did make it to Yosemite, it was right after high school, on vacation with my parents. Setting out from San Francisco, we drove for hours through parched hills and cemetery grids of almond trees fed on drip-tape. Eventually the landscape began to sprout and buckle and we ascended into a world of blue lakes and pitch pine. We camped the first night in the valley, and I sat outside for a long time in the dark, tracing the black rim of the cliffs against the sky. I loved being there, though buried in this love was a feeling almost like guilt, born of the awareness that I was enjoying an island of purity in a compromised country. Like all islands, its magnificence derived at least partially from its isolation and scarcity. It made my awe feel like an argument, a position I had to defend even though I knew it hinged on an exception to the rule.

The next morning, we took a stroll along the crowded paths that wound through the valley floor. Stopping to read all the interpretive signs, I noticed that many seemed to follow the same basic rubric. There would be a paragraph or two describing some natural process or phenomenon—the life cycle of a redwood, for example, or the hibernation strategies of a chipmunk—and then, in the final sentence, the text would make a brief reference to how the Problem was changing or jeopardizing the thing you'd just learned about. They were like little cliff-hangers, the way they got you invested in a particular wildflower or a specific vernal pond, then drew a big question mark over its future.

The problem with these signs was that you had to just take them on faith. I had no idea which ponds were supposed to go where or when exactly the purple loosestrife was supposed to bloom. Absent the placards that directed us where to stand and what to look at, nothing at all would have seemed amiss. I imagine this was true for most visitors to Yosemite, tourists who might come for one week every few years from whichever city or suburb they called home. ("It's our little escape this year," they might say.) The brevity of vacation precluded the kind of deep familiarity that would have allowed us to sense—without having to be told—that the baselines were shifting, the rhythms a little off. The very thing that made Yosemite beautiful—that it was literally exceptional, an ecological island distinct from our accustomed landscapes—made us more or less blind to its undercurrent of change.

Walking along the paths, I tried to notice some of the specific disturbances mentioned on the signs, but I kept distractedly craning my neck to stare at the iconic cliffs through the redwoods, awestruck by the panorama I'd seen so many times in paintings and documentaries, all those timeless Ansel Adams prints. It was magnificent, of course, better even than I'd imagined, but even so I felt a tug underneath it all, like the postcard vistas were in some subtle way coming between me and the actual landscape, like I couldn't see the trees for the forest.

The anthropologist Anna Tsing has suggested that in order to go on living in the Anthropocene, we will need to resurrect what she calls the "arts of noticing": the ability to stand still and just look, to pay attention without expectation, to welcome bemusement and surprise. To see the trees, in other words. Indeed, she focuses particularly on the value of noticing plants and animals. "There are other organisms that are key parts of our lives," she says, "and they don't always behave like resources."

Her argument is that these arts of noticing are particularly endangered within the system that created the Problem. They are accretive, not extractive; they prefer specificity over generalization; they require ample, unscheduled time. Reclaiming these arts is thus a means of rerouting the status quo, which might otherwise kill us all, humans and nonhumans alike.

I love and agree with this point, but there is also a painful catch. It puts me in mind of that scene from *Swann's Way* (which, yes, my therapist's allusion had dragooned me into reading) where the narrator is enthralled by a hedge of pink hawthorns. For several pages he stops in front of them, taking in every detail of fragrance and light, comparing them to beds of strawberries and silk bodices, to candy and cream cheese and the altars of churches. He forms his fingers into a frame so that all he can see are their blossoms, marveling that "it was in no artificial manner, by no device of human construction, that the festal intention of these flowers was revealed, but that it was Nature herself who had spontaneously expressed it."

But the arts of noticing have changed in the Anthropocene, and herein lies the catch. True noticing, as the signs at Yosemite teach us, will now unearth not just profusion but loss. In each late-blooming loosestrife there is a trace of every plane journey we've taken, every vote we've cast, every year that's been wasted to gridlock. The festal intention of flowers—and, at the very least, the timing of their expression—can now be said to reveal itself at least partially through devices of human construction. And so, just at the moment when we need it most, the inherent joy of noticing has been compromised, marred with mourning and with mirrors.

All told, our family spent only a couple days on the valley floor. Before arriving we'd secured a permit to hike the Yosemite backcountry, and on the third day we set off with our packs to escape the crowds (that is: to escape their escape). We pitched our first camp next to a clear lake ringed in granite. In the hour before dinner, I decided to take a walk by myself, wandering away from the lake into a maze of glacial erratics. The air was cold that afternoon and there was no wind, though the warped pines bore the mark of its latent ferocity. Except for the far-off gush of a waterfall, the scene was completely quiet. I was the only thing I could see that was moving.

From a distance I spotted a small promontory and scrambled up it to get a better look around. When I crested the ridge, there they were: a large brown bear and her jet-black cub. The bears seemed impossible in that inanimate landscape, like two boulders had come to life and were scuffling around in the dirt ten yards away. Slowly, the mother swung her massive head around and looked at me, seeming to focus on a spot just behind my head. For several long seconds I waited there, limbs frozen, eyes swimming in my head. The bear padded toward me, then away, nosing its cub and shifting its ponderous bulk. Then, as if arriving at some conclusion, it lumbered off to ransack some bushes.

For the next hour, I sat very still, watching the bear-boulders move across the granite. It took several minutes for the blood to stop threshing my veins, but eventually the adrenaline subsided and something weirder took its place: I began to feel that, in some important way, the bear in front of me wasn't real. After all, I had had to take a cross-country plane, then drive several hours, then buy a special permit, then hike for miles into a legally preserved relic of an old world just to see it here, in its dioramic context, amongst the rivers and peaks. The whole encounter felt so rare, so gloriously unrepresentative, that something in me struggled to lend it the full weight of an actual event. It felt instead like a flashback, as if the bear were a piece of the past. It was not unlike a feeling expressed by the narrator of *Swann's Way*. Toward the end of the book, he admits that hawthorns have become unreal to him, ambered away in the depths of his memory. "The flowers that people show me nowadays," he laments, "never seem to me to be true flowers."

Eventually, the bear that did not seem to me to be a true bear wandered into some brush and disappeared out of sight. I returned the next morning to look for prints, but the granite betrayed no evidence that either of us had ever been there.

After I saw the bear in Yosemite, I remembered another bear, one that was long dead and had spent its whole life in New Jersey. In the northeast corner of the state, if you know where to look, you can still find the safari theme park where this bear used to live, now abandoned and disintegrating in the middle of some scraggly woods. In high school, my friends and I spent several weekends exploring it, sneaking in through an unassuming tract of scrub oak and emerging onto an old asphalt path, which had eroded down into a trail of black crumbs. Nothing about the zoo had been preserved—not the forests razed to build it or the buildings themselves once they'd started to collapse. Everywhere you looked, former enclosures were being resorbed by vengeful foliage, ivy perforating their windows and rainwater warping their roofs. There were moldering plywood concession stands and a chain-link bird dome that looked like a giant, ragged basket. There was also an old tiger pit surrounded by a concrete amphitheater, and you could clamber down into it and prowl around like you were one of them, reclining on the fiberglass stones.

We'd all heard stories about the park's bankruptcy, how it'd been so sudden and complete that they'd run out of money to maintain the cages, and how for years afterward there were rumors of escaped animals wandering the second growth: kangaroos and ostriches and even, most notably, a large brown bear.

It was this bear that I recalled that day in Yosemite. I still picture it sometimes, roaming the fractal at night, tipping over trash cans, lumbering through parking lots. To me, this is the realer bear—the one I never saw, the one that may never have existed; the bear that escaped from the ruins of a simulated wilderness built on the ruins of an actual wilderness somewhere in the vicinity of the town where I grew up.

This is the bear that feels like the future to me, that I've retained as a private portent. Because in my head there is a scenario, either best case or worst: that eventually, over centuries and centuries, the Problem erodes all enclosures both legal and concrete. Everything escapes and everything comes flooding in. Yosemite gets ripped open like a birdcage. And when we've finally relinquished our pedestal and joined our fellow species in exile and witnessed the last of our Waldens resorbed into a land that is now neither fully wild nor fully tamed, then and only then do I imagine we'll realize what we've needed: an ability to notice the world that comes next. An ability, that is, to live there.

———

WHAT WE HAVEN'T SPOKEN about yet is the third category, neither human nor non-; that silent majority that eschews life altogether. I want you to remember this, too: that beyond people and creatures, there is still an entire world of stuff, a universe of objects that do not go extinct but simply sit there, dense with themselves, biding time.

In my job with NY Renews, and especially during those twelve-hour days I spent bent toward a screen in my office, I used to think about all of the objects in my bedroom sitting quietly just where I'd left them: a lamp, a pot, a rug—everything mindless and extant behind the closed door. When the dread began to rise—in those moments when the Problem felt impossible, claustrophobic, and foregone—I took comfort in the thought of this world beyond its reach, beyond any human concern really, a world that continued to exist in every room I left. My trick for regaining calm was to picture returning home to a certain corner of a bookshelf, or a specific patch on my quilt, anything that at that very moment I knew to be gathering dust in my empty house.

Then, once the endless day had ended, I would sit with the imagined object and examine it closely. If I'd chosen the lowest level on the bookshelf, I'd lie on my stomach with my chin in my hands and trace the plasticine grain of the fake wood, the whorls of brown muffled by dust. I looked in particular at where the shelf met the side, an angle slightly off right, cleaved by a thin gap of shadow where the pieces didn't quite fit. I would run my eyes along this gap, then sideways across the crenellated line where the bottoms of the books met the bare shelf, jumping forward in the hardcovers and then back again in the soft. At last I'd scoot away and take in the whole shelf, craning up at it in silence. I could spend half an hour this way, just looking at the thing, soaking in its aura of permanence and unconcern.

In retrospect, maybe this was a counterweight to the particular brand of solipsism that inhered in the Problem. This feeling of walking around all day with the end of everything slung heavily around your neck; the desire—irresponsible though it was—to lighten the load through recourse to that old conflation of world and self. At least then no one else was implicated. If it was just me out here, just my mind, then the Problem could be on my shoulders alone—a terrible weight to be sure, but still bounded, strictly personal. This was a temptation I often felt, to mitigate the cataclysm by undermining the reality of the world it threatened. (And was this not how the Pruitts did it too? Discounting the very *reality* of the worlds they destroyed, less a political attack than an ontological one?) Thank god I could never keep it up for long. How could I, when even from afar I could feel the static pulse of the objects in my room, objects whose very magic arose from their unflappable existence beyond me?

One of these objects was an old wooden desk, and in its smallest compartment I kept a booklet of postcards called "A World of Things." Each postcard showed one image from a series of woodblock prints produced by the Japanese artist Kamisaka

Sekka. Sekka was known as one of the last great masters of tra-
ditional woodblock printing, though he departed from his predecessors' preference for detail, opting instead for simplified
shapes puzzled together in bright contrasts. I liked his prints
because they reminded me of my favorite haikus, which tried to
locate an instant of time in the body of an object. One postcard
showed a pair of abstracted fans placed next to each other on
a table, another a thicket of pine trees, each trunk a simple stroke
of brown. Everything in his world of things—living and dead,
human and non-human—seemed to thrum with the same soft
frequency. In one postcard I remember, a boy leaned against the
flank of an ox, playing his flute. The image paid no preferen-
tial attention to boy, flute, or ox, depicting them all with the same
still-life flattened weight.

My favorite of the postcards was titled *A Traveler*. It showed
a man's head and shoulders emerging from a pass between two
hills. The man himself was grimacing, though the enormous
rucksack on his back looked slyly content, like it had hitched a
free ride to a destination of its own. At the crest of the pass, a
patch of red flowers awaited them, privately gloating, at least it
seemed to me, over having been the first to arrive. The rest of
the image was nothing, just tan air and dark gray hills.

For years I wrote all my thank-you notes and birthday wishes
on these cards, slowly disseminating Sekka's strange trees and
snowbound-huts to friends and family. I never wrote them anywhere other than my apartment, but I liked to think of them as
actual postcards, dispatches from a foreign place I was visiting.
Greetings from the world of things, they seemed to say. Wish you
were here.

I was never successful in actually visiting that world, though
whenever I looked at one of Sekka's postcards I would try for a

moment to summon the probably impossible conviction that I was just another one of the things in my room—voluble and mobile perhaps, but fundamentally of the same category as the lamp and the desk chair. This habit was motivated by yearning—for permanence, or maybe just stillness—but it was also driven by a weird sense of obligation.

What separated the Anthropocene from the Holocene was that Homo sapiens had invested itself with geologic power. People were shifting from subjects into objects, graduating into an ontological category populated by mountains, oceans, and ozone—everything as powerful, implacable, and ubiquitous as we had lately become. It seemed to me that we had a duty to contend with our newfound status. Even small inroads of understanding might help us inhabit it more responsibly. So I tried in vain to assume this new role, to become an *it* and see how it felt.

For a few weeks most summers, my family would rent a house in Maine. The beach near this house was fringed with hard stones, which tumbled against one another in the drag of the waves. My father used to take me down to examine them, how each was uniquely webbed and stippled, rounded into the shape of an egg or a shingle. He'd pick one up and turn it over in his hands and then hand it to me gently, as if it might break. Most were granite cut with quartz, though some held traces of iron red or purple, even green. They were extremely old, he explained, hundreds of millions of years old, much older than our species, older even than the ocean they bordered. I'd hold them with reverence, thinking that I could feel the time they carried in their weight, which was always slightly heavier than expected. Our predecessors and successors, I called them; our once and future kings.

Sometimes we'd balance them into towers and come back the

next day to find them still standing, framed against the sea. Or they had collapsed and were scattered about, with rarely a chip to indicate they'd ever occupied positions other than the ones they'd fallen into.

At some point in my twenties, I took back a stone—a particularly beautiful one, smooth dark gray with a golden band of quartz—and placed it on the little table in front of where I meditated. For a time, it was like a miracle every time I saw it there. To think that it might stay that way, exactly as it was, forever. A little glob of eternity perched inside my mortal home.

The writer Jenny Offill, whose own work grapples seriously with the Problem, captures a similar dynamic in her novel *Last Things*. Partway through the book, the eight-year-old narrator considers the decorative stones in her family's garden: "My mother said that stones were last things and would be around long after people were gone. Other last things were oceans, metal, and crows. I thought that if I filled a birdbath with seawater and dropped a coin in it, I might glimpse the end of the world."

Maybe this had been my secret hope, too, when I brought the stone home. That it would afford me a glimpse, act as a kind of window into the future. Maybe this is why I placed it in front of my meditation cushion, so that the last thing would always be the first thing I saw when I opened my eyes, though ultimately I found that even in those moments of literal wakening, the stone never really looked like a window into anything—just an inscrutable granite lump, almost anatine in its self-possession.

I began to think of the rock as more of an envoy, not something that would allow me to see the end of the world, but something that would be there to see it for me, when the time came. ("Last" being, of course, not only an adjective but a verb, the critical thing it could do that I couldn't.) During the fifteen minutes in which I meditated each morning, I noticed myself trying—with a desire that was almost envy—to become more rocklike. I would sit there facing the stone with my shoulders locked, willing myself to adopt its countenance, petrifying anything that came into my head. This quickly came to feel like an embarrassing sort of prostration, like the stone was a god I'd placed on a shrine. Though in a way, it turns out, this wasn't entirely wrong.

Because by now I think we can acknowledge a further truth: that coal is a last thing, too. And carbon and mud slides and flame. And all the oil sluicing through all the pipelines that in my head I've pictured destroying. Everything we've consigned to the world of things, everything we've presumed dead, is revealing itself now to be in possession of a terrible agency, a higher-order control beyond what we ever could have imagined. Their inanimateness has been little more than a clever veil. They are autocrats at base, incapable of negotiation. The crust and the sky; the sludge, sea, and dust: these are and always have been the true subjects. It is we who are the objects now, intermediaries for their massive churn. And as we dig them up, they bury us—so slowly we barely notice.

How to understand the role of objects in the Anthropocene? First of all, they are us. With the advent of humanity as a geologic force, our species has undergone at once a profound humbling and a striking promotion. Objects are our kin now—or rather, we are theirs—which means we have to understand ourselves as invested with the same colossal powers and responsibilities as an aquifer or a mountain range, capable of shaping the inorganic circuitry of the planet itself.

Yet even as they've become our kin, they've revealed themselves also to be our destroyers, enacting vengeance for centuries of betrayal. Though we should not ascribe to them any human intentions—it's enough to concede that they have agency: dispassionate, unaccountable, undeniable power over us. The fear now is that our final undertakers will emerge from the world of things: the oceans, fires, and photons come to repay our indifference, driven not by malice but by the far more implacable fury of physics.

But there's a third side to objects, I think, one that is quieter and less obvious. It requires we see them not as inescapable, but as their own kind of escape. Because even as they augur the end of the world, they are each of them harboring another one inside themselves, a refuge which, with enough careful attention, can be unfolded from any piece of furniture, any island in a parking lot.

There is a writer I often return to named Gerald Murnane, a man who has never left Australia and barely ventured beyond his home state of Victoria. He claims never to have been on an airplane, never to have used a computer, never to have learned to swim, and never to have set foot in an art gallery. He lives by himself in a bunkerlike home in a drought-prone town in the outback, crafting stories with a single finger on his typewriter.

Yet out of this sparse ground his prose tills endless meaning, whole cosmologies formed out of the divots in a certain hill, or the memory of an image on a postcard, or the subtle variations in color among the racing silks worn by the jockeys in the local horse races. As Murnane explains in his novel *The Plains*, the inhabitants of his fiction are engaged in the "lifelong task of shaping from uneventful days in a flat landscape the substance of myth." They examine their monotonous world as if it were exploding with detail and possibility, as indeed it is when observed through Murnane's prism. His short stories in particular have an obsessive quality to them. He retreads the same raw material over and over again, finding new layers of resonance and association until he's built—from almost nothing: a few remembered objects, the distribution of grass in a yard—a seemingly infinite web of signification, a map of his entire mind. The result is both prosaic and miraculous, like the everyday emergence of consciousness from tissue.

I guess what I'm trying to convey to you is the genuine hope this gives me. Or, if not hope exactly, then at least a sense of lateral possibility: the thrill I feel in witnessing the near-limitless human capacity to make things matter. Any things, whatever is given. It's a form of power, I think, the source of our mettle. How can the world ever end, if a new one can always be summoned from the hottest, most barren country? This is the feeling: that even if the Problem boils everything down, dries the lakes and cracks their mud, leaves things as empty and featureless as the inner craton of Murnane's imagined Australia, still. What's left could suffice a world, a million.

I wouldn't blame you for thinking that this is all just another retreat. That as the animate world collapses, I'm simply falling back on the inanimate, taking solace in its endurances. But this sort of passivity is a dead end, the last thing I'd want to leave to you. So maybe I should be clearer on this point: I am not advocating that we withdraw to the world of things, living out our days in their fatalistic contemplation; I am not advocating that we forsake our responsibility to vulnerable people by seeking comfort in the eternity of objects. The Problem is still a force of enormous, politicized, and asymmetric violence, and it is still within our power to allay it.

What I *am* suggesting is a new kind of noticing, a communion with the last things—not as an excuse to stop fighting, but precisely as a means to continue. To always take heart in what's left, to lavish it recklessly with meaning, to grab grit from the dirt that outlasts us.

Like a salamander losing itself among stones, this is not a form of forfeit. It's an act of subterfuge, of imagination. In a world punched with holes and lit by flame, it is a means of staying alive.

Fourth Movement

After the first few hours of walking, my soles become acquainted with the different types of sand. There is sand so soft and so dry that you sink to the ankle on every step. There is sand whose top layer has caked to a crust, so that each footfall cracks the surface into floes that can be picked up and crumbled in the hands. In places the crash of waves has made the sand as hard as cement, and the pads of your feet throb slightly in its crossing. In other places the sand is only half wet, and the weight of your foot presses the moisture outward, creating little parched perimeters, like paving stones that appear as you walk. Sometimes the dynamics of this saturation are so complex and inscrutable that your stride can effect a tautening even in sand that is several meters away, so that as you walk you notice little bursts of light up ahead of you, wherever the water is wicking through the grains.

In the mornings, the sand is cool and fine as talc, but by midday it's grown too scorching to bear, and we walk only along the lower beach, where the ocean proffers its fans, which sweep across our shins and scrub away our tracks. Around sunset we retreat once again, back up the beach, to where we can watch the skeins of water reflect colors we've never seen. Oranges with the bite of metal, irradiated reds. And in the moment after the sun disappears but before its light has fully left the sky—when the palette above the horizon mimics exactly the palette below and the radiance along its seam fades outward into the blackness of night and of dirt—it is possible to experience that old hackneyed sensation also induced by the surfaces of ponds and by particularly vivid dreams. Specifically, that you do not live in the world, but a version of it, and that behind its mirror there is another, of equal weight, endowed also with the power of looking.

At night we set our camp in the thicket behind the dunes. Of the several dozen people in our party, most are members of the Goolarabooloo clan, the Aboriginal family who for generations have acted as custodians over a territory that extends north from Broome—a small, coastal town in this remote region of northwest Australia. Four generations have pitched their tents here among the scrub—toddlers and teenagers and greatgrandparents—all of them gathered in fulfillment of their ongoing responsibility to maintain relations with the land. I'm here as part of a contingent of outsiders: non-Indigenous, mostly white people, who've been invited to join the Goolarabooloo in their perennial movement across this stretch of country.

One of the fathers of the clan—a man in his late forties who shares my first name—has been charged with shepherding this latter group. Around the campfire, Daniel tells us stories of another world, the Dreaming, which enfolds our own like a shroud. In this other world, everything is both itself and something else. Twin spits of rock are the fangs of giant serpents, and dry streambeds are the furrows left by their windings. The boulders we saw lining one section of beach are also the chaff produced when winnowing grass seed from stone. Three paperbark trees—out of place, set apart from any grove—are the digging sticks left over by three sisters, who have themselves become three pillars of rock farther down the coast.

This correspondence extends all the way up to the sky, where Daniel points out a dark patch between Scorpio and the Southern Cross, a kind of anti-constellation. It is, he shows us, shaped unmistakably like an emu. In the cosmology of the Dreaming, the emu spirit helped create the world, traversing its features and singing them into being. When at last he grew satisfied, he stopped roaming and launched himself into the sky. The story goes that he remains there today, watching over his work, heralding the passage of the seasons.

Which is to say that absolutely nothing here is without agency—even emptiness itself, the void between stars, can harbor enormous consequence, the hidden germ of an entire life.

Every year the family walks this land twice, attending to it, keeping its company. They camp in the same spots, visit the same springs, scour for honey in the same paperbark trees. The journey is about sixty miles in total, though they never rush it, stretching the trek out into a stroll over eight, sometimes nine days. Were Thoreau to have made the same journey, I imagine that in his blindness he would have seen a wilderness here, a beautiful and mostly empty place in which to horde his solitude.

Indeed it is beautiful, this particular arc of coast. There are white dunes abutting cliffs the color of blood. There are dolphins and rays that leap up in glimpses, and sharks that skim by on the prowl. Behind the dunes there are vast, muffled tracts of acacia and gum, vine-draped groves that spread inland for miles, their silence broken only by the far-off throb of the waves. Few people live here, but the landscape is not empty. Wherever you turn there's the thread of a story, the sight of a spirit. The country is not wilderness at all but crowded, dense with characters and plots, dangers and havens, pre- and proscriptions laid down over millennia.

Connecting all of this is a physical trail, a path through the landscape, variously referred to as a heritage trail, a Dreaming track, or a song line. To follow the trail is to plot the land, to replenish it with narrative. And so, every year, all the aunties and the cousins and the little grandbabies of the Goolarabooloo clan come out here to give time and to pay attention and to prevent the two worlds collapsing into one.

Years ago, the fossil fuel conglomerate Woodside Energy had designs on this coastline, wanted to build a gas plant here that would have processed fuel from the rigs offshore and shipped it off again to markets in Asia. The plant, it was rumored, would take up two square miles of land, a region of red sand and gray brush right in the middle of the song line. The state premier, who made no bones about his support for the project, referred to the site as an "unremarkable stretch of country," and went on television to stump for its development. When the clan saw this, they knew a collapse was at hand, that their country was being transformed, impossibly, into a grid of numbered parcels, and that after the gas plant there would be no more plotlines, only pipelines, so thick you couldn't see the sand underneath them.

So they fought, enlisting allies from around the country and establishing a protest camp adjacent to the site. Local businesses started shipping over food and supplies. Builders loaned their equipment. A crew of citizen scientists set up a survey platform to monitor the humpback whale migrations just offshore, gathering data that plainly contradicted Woodside's rosy environmental impact statements. The standoff went on for months, then years. For long stretches the protest camp had over a hundred residents, a rotating cast of journalists, academics, drifters, locals, and activists from way off in Sydney and Melbourne. Many of the non-Aboriginal people who came out were those who in years past had taken part in a walk of the song line. For the Goolarabooloo elders presiding over the camp, all this was as planned. Inviting outsiders to the trail had never been about tourism. All those years, they'd been building a constituency.

Eventually the protest caught the attention of some larger organizations: the Australian Green Party, the Wilderness Society, the Sea Shepherd. Rallies were held across Australia, five thousand people gathering in Melbourne, twenty thousand in Perth. As the project delays lengthened, the Woodside security personnel grew more aggressive. They'd show up to camp in the middle of the night, shining flashlights into tents and pointing video cameras in peoples' faces, their name badges blacked out with tape. The protesters, too, started escalating their tactics, chaining themselves to bulldozers and drill rigs. Finally, in 2013, Woodside threw in the towel. Company executives laid the blame on falling gas prices, but many of their joint venture partners— fossil fuel goliaths like Chevron and Shell—had already pulled out. They'd expected free rein in this obscure and "unremarkable" country; they'd expected it to be empty.

Years after the fact, I was sent an article about the Goolarabooloo's victory. It quoted Daniel's uncle, Joseph, who'd emerged as a de facto leader of the anti-gas campaign, and had spent years dashing from rally to rally. He wanted his children to inherit a connection to the country, he said, and pitied the Aboriginal people who hadn't. "They walk around with a dead feeling, these people, inside them."

This indictment stayed with me, replaying in my head as I began to write this letter. I decided to apply for a fellowship to an Australian university. My proposal was to finish the letter and to walk the trail with the Goolarabooloo. The two would complement each other, I argued. If I could understand how the family passed meaning on to its children even under threat of a collapse, then perhaps I could learn to do something of the same. When I found out, months later, that I'd been awarded the money, my gratitude came with an undercurrent of guilt. Not only due to the fact—how had it not registered?—that I would now need to take two carbon-spewing flights across the Pacific, exacerbating the Problem I'd intended to address. But also because I suspected, deep down, that my motivations were more selfish than the ones I'd articulated to the application committee. Despite my criticisms of Thoreau, I think a part of me was still looking for that untrammeled glade, a far-off place that had been spared the pipelines. As if this alone might offer a respite from the Problem. As if it were still possible, in the Latourian sense, to step outside.

The Goolarabooloo's victory came at great cost. There are other Aboriginal people in town who still won't talk to the family, who were holding out on the promise of jobs. Joseph himself died at the age of forty-seven, struck down by a stress-induced heart attack ten months before the conglomerate withdrew. In Aboriginal culture it is taboo to speak the names of the recently deceased, so I never hear his name said aloud by any member of the family, learning it only from a printed obituary. Even when they recount the long history of the campaign, they mention Joseph only by implication, skirting around him like a hole in the narrative.

Despite this, there are no regrets. They have seen what's happened on the Burrup Peninsula, just down the coast, where the gas industry got a foothold and never let go. I have been there myself, and I can tell you the contrast is terrifying. The peninsula has become a piece of infrastructure, like the hydraulic arm of some immense machine, its flatlands obscured by mazes of pipe and industrial wharves and storage domes the size of the surrounding hills. If you climb up one of these hills, the feeling looking down is like staring into Mordor itself, a vast, clanging city, complete with a giant flare stack and its perpetual eye of flame.

Most areas of the peninsula are closed to the public now, requisitioned behind razor wire and guards' huts, though in the middle there is a tiny national park squeezed between the fences. On the sides of the red boulders that line the ridges in this park, there are rock carvings which have been dated back thirty thousand years, to when the seas were lower and the islands offshore were just farther hills. The carvings are faint, but if you look closely you can make out emu tracks and boomerangs, ghostly people with long arms and domed heads. Many of the rocks are engraved with symbols whose meanings have been lost, though when I visited I would from time to time spot a carving whose candor and familiarity took my breath away. Pecked into the underside of one boulder, for instance, I found the perfect, sinuous figure of a dolphin, not an ideogram but a portrait, as if the artist had caught it midleap. It is said also that the carvings on the peninsula include the oldest known representations of the human face, though it is unclear how many of these have been bulldozed by the gas plants, which is only part of what I mean when I say the landscape has been defaced.

Archaeologists and anthropologists generally consider Aboriginal culture to be the longest-running continuous culture on the planet. According to most estimates it is sixty thousand years old, give or take. To get a sense of just how old this is, consider that only fifteen thousand years ago, the land that is now New York—that triumph of towers and tunnels, the city that never sleeps—was buried in darkness under a mile of ice. So when Daniel and his relatives invoke the holy sentience of the land to ward off the corporations that would demolish it, they are drawing on a cosmology whose sheer longevity makes the birth of Christ look recent.

Given all this talk of prehistory, it is often wrongly assumed that the Dreaming is located somewhere in the deep past, a period so distant it can now be safely populated with myth. But if you ask the Goolarabooloo clan, they will tell you that the Dreaming is ongoing, that it never ended—nor could it, really, because what would it mean for the land to suddenly lose its train of thought? When exactly was the moment when the water and the rocks and the snakes and the storms relinquished all agency, when they stopped telling their own stories? The whole premise is absurd, that we could consider ourselves so detached from the world on whose umbilical grace we so palpably rely that it would actually become possible for us to drill into it without expecting a reply. It is frankly naive, in their view, to think we live "in an environment." To Daniel it's much more like a society of actors, each with its own influence and intent, its own private dream of the others. And it is this Dreaming that he says cannot stop, that is neither ahead of us nor behind. The anthropologist William Stanner has argued that the Dreaming occurs in the "everywhen," a time so fluid and capacious that sequence fades toward irrelevance, a minor detail impeding comprehension. Which is to say that there are times much thicker than my sought-after Now, with its constant arrivals and departures, its flitting brevity. There are times so big that they might actually contain everything.

In the West we taxonomize time, cleaving the Anthropocene from the Holocene, late modernity from the Renaissance. This lends the illusion of beginnings and endings, recalls junctures where there was only ever flux. The practice facilitates a fantasy if not of control then at least of separability: that certain phases can be entered, others left behind. The everywhen contests this delusion, sweeping the past into the future and vice versa. It allows us no escape from the Problem, but neither does it consign us to an outcome. It just holds us here, in the Dreaming, never less than fully implicated, never more than partly in control.

Our second day on the trail, I approach Daniel midstride and start asking questions about the Dreaming. He is wearing a baseball cap and wraparound sunglasses, and his flip-flops slap as he walks. I ask him about the meanings of certain rocks, the uses of various shrubs; he lets me prattle on for a few minutes before gently rebuffing me. He's noticed, he says, that the white-fellas like to have a chat as they walk. It's a way to pass the time, after all. But sometimes silence is the best thing, he thinks, it lets you pay closer attention to the country you're walking, lets you notice things you wouldn't otherwise. Chastened, I try my best to keep quiet, and turn outward to the business of noticing. I stare dutifully at many things but can only attend to some of them. I notice, for instance, a wattle tree whose yellow flowers look as thin and bristled as pipe cleaners. I notice another with white flowers, fatter and more silken, like skeins of yarn. I notice that the shadows which sometimes cross our own belong to two distinct kinds of raptor, and that if you really scan the sand you can spot the prints of dozens of animals, among them the hermit crab, whose tracks look like the tread of a tiny bicycle tire. I do not notice the small holes in the paperbarks that indicate the presence of honeybees. I do not notice, until one of the aunties points it out to me, that the tide sometimes strands minnows in the little lagoons, and that these can be gathered up easily for bait.

My meditation practice—all those inexpert hours spent looking inward—have left me ill trained for looking out. Those few things I do manage to observe, I find myself needing to share with someone, as if to set their beauty down in record. Look at those cliffs, I say; look at that bird. Look at how the sun catches the water like a net. Eventually the beautiful things become too numerous to name, and I shut up about them, letting them sit there in my head, gradually shedding their metaphors.

There are many children on the trail, grandchildren and great-grandchildren of the Goolarabooloo clan, and most of them wear no shoes. They chase after one another over hot sand and sharp rocks, then get tired and walk for hours in the shell of their thoughts. But though I try to pay attention, I never once see them examining the ground in front of them, deliberating over where to step. Their soles are thick, and they seem to look only ahead, or around, or at nothing. My own feet get cut as I walk, and the cuts pack with sand, but it's an easy thing to walk down the beach and let the water wash them clean.

The next day the group comes to a brackish creek with wide banks and a slow current. The creek is too deep to cross by wading—and is anyway home to large crocodiles—so we wait for a metal dinghy to ferry us the short distance across. While we wait, Daniel tells us about the tides, which in recent years have grown significantly higher, spilling over the sides of the creek and wearing away at its banks. The flooding got so bad one year that the rising water washed out a sandbar containing an important ancestral burial site, dissolving it in a matter of weeks. Later, some white tourists happened upon the bones, clearly human, scattered here and there along the beach. Daniel reckons that they could've been a few thousand years old, though the analysis was never done and he has no way to confirm this. In his explanation of these events he speaks about the Problem at length, and I'm reminded that the hyperobject exists even here; that it can infiltrate any place, exhume any past. That whatever exempted Eden I'd pictured vaguely in my head, it had always looked less like Broome than like Walden—a fantasy of purity sustainable only through the withdrawal of attention.

Months later, the last vestiges of this fantasy would be swept away. The hyperobject would explode into bushfires so vast and ferocious they'd eventually destroy an area the size of England. The fires wouldn't start anywhere in particular, they'd just erupt everywhere all at once, like hives out of skin. They'd reach from the forests of the east coast to the scrublands outside of Broome. I'd end up flying out of Australia on a windy afternoon, with temperatures above one hundred degrees Fahrenheit. Days later, the fires would reach the hills overlooking the airport.

Later, back in the States, a friend would call and tell me about driving home from Sydney to Adelaide for Christmas, a trip of about fifteen hours, the same as the distance between New York and Atlanta. He drove through smoke the whole way, he told me, peering through it out his windshield, tracking its movements on the radio. Always there were fires both ahead of him and behind, like he was being followed and anticipated, like he was being chased. Sometimes he saw them licking up behind nearby ridges, flames seething under a weakened sun. When he finally arrived in Adelaide, he joined some friends on a nearby island, famous for its empty beaches and wild platypuses, until even this haven started to smolder, then burn, forcing its residents to evacuate.

This was what it felt like to live "inside." To drive for hours and never emerge, to have the hyperobject move with you. Even the Goolarabooloo, when they blocked the gas plant, had not escaped its bounds. Their land was still subject to the Problem, it'd just been spared the fate of contributing to it.

The fact that I'd ever imagined the Goolarabooloo were exempt was symptomatic of a larger issue: the fantasy of purity extending from the land to its people. The late Aboriginal economist Tracker Tilmouth used to deride this instinct, saying that some environmentalists preferred to see indigenous communities as forever static and pristine. Indeed, I would sometimes notice the white people on the trail, myself included, talking about the Goolarabooloo as if they were saints or saviors, emissaries from the past come to deliver some immutable piece of wisdom. People would speak in hushed tones about their humility, their alertness, their skill with a fishing spear. None of these things were untrue, necessarily, but the romanticization was its own kind of stereotyping, and anyway didn't admit the whole story.

Out on the trail, the Goolarabooloo did not conform to any fantasy of themselves: they ate cup noodles and smoked menthols and drove giant pickup trucks across the beach. Many members of the family didn't physically walk the track at all, instead piling into the beds of the trucks and getting to the next camp early, so they could spend the whole day fishing. On one occasion, I hitched a ride down the beach with an uncle who was sporting a Kobe Bryant jersey and a mullet, and whose jeep had no windows or license plates. I had to shove past a cooler of beer to get in, the empties rattling around at my feet. He was a little drunk, it was clear, but nothing serious, and the beach presented no obvious obstacles, though I wasn't surprised when we drove over a branch and popped a flat. There was no spare in the trunk either, so we just inflated the still-perforated tire as best we could and sped to camp as the air hissed back out the hole, the entire chassis listing more and more to the left. All of which is to say that, between the two worlds, it is possible for a trip down the trail to be at once a spiritual rite and a family vacation, and that the messiness and petty comforts of the latter do not negate the sincerity of the former, the genuine will to spend time with the country.

I once saw the Aboriginal writer Alexis Wright speak at a symposium on "Writing in the Anthropocene." In matters relating to the Problem, she said, "It's become a bit of a fashion to talk about indigenous knowledge, to try and package it." Of course, she went on, what is often referred to as the indigenous worldview does have much to offer us on this front: concepts of interdependence and nonlinear time and nonhuman sanctity, a "long vision" that can learn from generations past and account for generations to come; all concepts once dismissed as idealistic, revealing themselves now—in a time of worsening floods and droughts—to be the most clear-eyed kind of pragmatism. It is also true, she said, that the Problem is not difficult for indigenous people to conceptualize. "It is easy for Aboriginal people to imagine that there would be greater populations oppressed and dispossessed, and unprecedented poverty creating even greater divisions in all of humanity. These are issues we have been dealing with as Aboriginal peoples for a long time."

However, she warned us, we had to remember that culture was never static, that it was always adapting to its time. For this reason, she'd begun referring to some of her novels as "survival literature," ongoing and increasingly urgent attempts to envision the fate of the world and the place of Aboriginal people in it. "I wondered how far we would go to retain our sovereignty of mind . . . how far we would go to continue believing in ourselves." The point being that—in spite of any half-baked hopes to the contrary—Aboriginal culture did not have some silver-bullet answer to the Problem. In fact, she told us, "I have no doubt that our struggle to survive will become far more enormous than the catastrophic realities of our current times." So she did not want to give the impression of offering easy reassurances packaged as fail-safe ancient wisdom. She was very clear on this point: "What I'm trying to deliver to you is something I'm grappling with myself."

Listening to Wright, the irony felt world-historical: that after centuries of colonization and exploitation, the European world was beginning to look for salvation in the minds of the very people it had dispossessed. There was no denying that Aboriginal civilization had survived sixty thousand years without giving rise to the Problem. What its defenders often referred to as "Western civilization," on the other hand, had managed to bring it about in just a few thousand—in part by robbing indigenous people of their lands and siphoning off the fuels underneath them. And herein lies the danger: if we in the West choose to mine indigenous culture for solutions the same way we mined indigenous land for carbon—engaging not with people, but with whatever idea of them best suits the program of our relentless prosperity—then any succor we find will be temporary, another trap we'll have sprung on ourselves. We cannot extract our answers any more than we can outsource our questions. Of her own writing, Wright likes to insist that she's "learning here too." And as any good teacher will tell you, real learning can only occur when you haven't already presupposed what it is you're going to learn.

I think about this often as I am walking the trail, how whenever I am looking for something, I never seem to notice it. True noticing involves an element of surprise, a vulnerability to the unexpected. Append a teleology and it's all over; you'll only see what you already have.

So I am not, in these pages, going to hand you a nostrum unearthed from some indigenous idyll. All I can offer are a few wrinkles in the picture, the things I noticed when I wasn't looking for them. For instance: Aboriginal people are often spoken about as having an emotional connection to their environment, and, indeed, Daniel himself speaks in exactly these terms. But the accompanying assumption—at least the one I myself once subscribed to—is that the emotion being described is always something akin to love or respect or harmony. The Goolarabooloo people I met do love their land; they love it fiercely, with a fealty that leaves me in awe. But they also sometimes fear it, or compete with it, or grow impatient with it. In camp at night, aunties would warn me about camping under certain trees. They're bad luck, those trees, they'd tell me, they'll bring you nothing but trouble. Sometimes we'd pass by whole sections of forest that the men weren't allowed to enter, or the women weren't allowed to enter,

or everyone generally was advised against. The spirits there were
too dangerous, we were told, and we listened. And it shouldn't
come as any sort of revelation to know that members of the clan
sometimes expressed frustration when the fish weren't biting,
and annoyance when the sun made them sweat, and mild boredom when the walks took too long. This was the relationship
to country I noticed. Not the romantic projection but something
much more profound: the full spectrum of feeling—the mundanity, the variety, the sheer detail—that comes with being part
of a family.

In one of the interviews he did for television during the contro-
versy, the state premier was asked about Joseph and the Goolarabooloo's opposition to the gas plant. I understand, he said in a
tone that indicated he did not, that Joseph has an *emotional* connection to that area, but we can't let that get in the way of what
the economy needs. Like someone stopping midkidnapping to
inform you that—regrettable as it is—the market value of your
family members simply necessitates their abduction.

   We can table for a moment the long-term economic mad-
ness of investing in an industry slated to crater your GDP in the
coming century. What was most revealing about the premier's
condescension was the fundamental difference it laid bare: not
between love and disdain for the country, but between any feeling and none at all. Between a family whose bond to the land
was emotional, and an official who refused to imagine what that
meant.

On the sixth day we reach our final camp, and on the seventh
day we rest. People lay out mats in the shade or drift off to

wander the bush. I accompany some of the younger cousins down
to a near-shore reef, where our plan is to spend the day fishing.
It's low tide when we get there and the reef has revealed its face,
a platform of greenish rock extending a hundred yards out to
sea. One of the cousins, a man about my age, wants to show me
how to cast a handline, and we spend half an hour practicing
how to hold the cheap plastic spool, how to flick your wrist so
the line doesn't catch. I'm useless at first but manage eventually
to land my hook a decent ways out, and he is satisfied enough
to head off on his own. I watch him stalk the edge of the rock
staring down into the ocean, an old T-shirt wrapped around his
head. In one hand he holds a lit cigarette and in the other a
wooden spear, and when he leans back to throw the spear he
squints his eyes and clamps the cigarette hard between his lips.
The rest of the cousins are spread out along the reef, fingers
testing their lines, which angle and glint into the surf. In two
hours, they've caught a dozen fish—mackerels and butterfish
and Spanish flags—unhooking them deftly and chucking them
into tide pools, where they continue to swim in tight circles. I
only ever catch rocks, but I don't mind; it is good just to be out
there, standing still in the spray from the waves. As it passes, the
afternoon leaves its tracks across my skin: the cut of the line, the
heat of the sun, the salt, which dries white on my arms. There
are delicious little oysters studded around the reef, and sometimes I pause to go crack one open, though the rock is so sharp
that even in sandals I am forced to walk very slowly, feeling the
shape of every step.

When we return to camp that night, some of the older men have
caught a sea turtle, and its shell sits like a giant shield over the

embers of the fire. As its meat is passed around, they regale us with the story of its capture. Apparently the uncle who spotted the turtle wasn't allowed to spear it since his wife is pregnant, so he had to call on his brother, who threw his spear so hard he pierced the shell in one go. There are all sorts of complicated taboos like this, about who can hunt what and when. Many people here have totem animals, creatures they consider to be close relatives and so abstain from killing. It has been suggested that this totemism functions as a mode of sustainable management, limiting the number of animals hunted by any one clan in any one season, though the Goolarabooloo talk less in terms of "management" and more in terms of relationships—relationships of rivalry and reciprocity, of giving and taking, relationships too knit to untangle. Whatever this country is, it is certainly not a wilderness, somewhere to be left alone and preserved in state. It's a place to engage with, to feel, to spear and stare and sing into being. Sometimes when I think about the incalculable damage being done by the Problem, I feel so guilty in my humanity that I'm tempted to recuse myself entirely, to bolster that arbitrary line between "people" and "nature" and then retreat behind its wall. But this is a fantasy of extinctionists and extractivists—that we could ever fully sever our connections, that it would even be possible to choose one world and forgo the other. As Daniel likes to say: "Country needs people." You can't withdraw when you have a role to play.

Still, if the conglomerate had won, it would have done its best to effect an untangling. This would have proceeded stepwise, through a series of abstractions. First the country would have been abstracted into soil, then the soil would have been abstracted into acres, then the acres would've been abstracted into a site, then the site—having been abstracted of all vegetation— would at last have been abstracted into a plant. Then the plant would have set about its business of abstracting fossils into fuel, which would in turn have been abstracted into markets, which would have done their job of abstracting the fuel into currency, which might afterward have been abstracted into derivatives or futures, so that by the end of the process the final traces of the original country would be contained only in tiny symbols on the screens of financiers, tiny molecules accruing invisibly in the atmosphere. One might be tempted to claim that the Dreaming is also an abstraction, that its stories of spirits and totems are abstruse if not incredible. But if this is true, then the same can be said of the Pruitts' stories, those superstitious fables of endless growth and divine right that Coates has called the Dream. And where the Dreaming grows its stories up from the land, the Dream paves the land over with its stories: narrow, inharmonious stories that must silence all others, that can only accommodate a single world.

And so we walk, against this silence, across the beach. We walk to remember the country, not what it might be but what it is, what it feels like. We walk because we want to notice, or maybe be noticed; to place our ears, eyes, noses, mouths, and skin to the ground, so it can speak to them, finally, as it will. Because it is harder to abstract stones that have cut your feet, harder to abstract fish that you've reeled in and eaten. None of this is a solution, of course: we cannot simply walk our way out of the Problem, nor can we expect the Goolarabooloo to carry us. But as your future deepens, maybe walking will prove itself to be a useful form of thought, an ontological habit worth cultivating. Just to tread lightly, without intent; to step through the symbols and into the sand.

On our last day of walking, we reach the country where the plant would have gone. The cliffs are tall here, and deep red, and there is a joy in standing under them, in looking up at their faces and knowing they've been salvaged. And this goes without saying, but the premier was wrong to call this country unremarkable: all around are the tracks of dinosaurs, hundreds of millions of years old, the ancient fossils of other walks. The largest of the footprints is nearly six feet across, the print of a sauropod thought to be the biggest animal ever to walk on land. It looks from afar like a giant tide pool set into the rock, and when we arrive Daniel is standing in the middle of it, smiling. The clan has known about these tracks for ages, and of late they've been collaborating with paleontologists to document their variety and extent. Had the conglomerate built here, they would likely all have been lost, and they may still disappear beneath the rising tide. But for now they are right in front of us, and so we step into them, plant our feet inside theirs.

Nearby there are smaller tracks, three-toed, distinct. To the paleontologists, they are the tracks of a theropod; to the conglomerate, the traces of fuel. These truths are true too, but also: witness the tracks of the emu, where his spirit stepped off and ascended.

Heat

At first there was just the numbness, like a hard shell. Inside it was the Problem, and inside the Problem there was you. I tried to break through but the walls were too thick. I'd spent years fortifying them, not least against myself.

So for a time I had to make do with the shell itself, its cold, smooth surface. It was a thing I could only observe through inference, the way physicists speak about dark matter—something detectable only by its absence, by the hole it leaves in the picture. But every absence has a form, and slowly I found I could trace its contours, could feel the weight of the shell pressing down behind my eyes and into the folds of my gut. It was so heavy I knew it had to contain you, though this felt like nothing, just one of the many mute truths that described the empty circle where some kind of grief was supposed to go.

Eventually, I started writing this letter. I wanted to explain the shape of the numbness, make some sense of it for you. But the more I examined its surface, the more I noticed the cracks—little fissures of feeling, smaller than the gap between words. I began writing into these cracks, trying to pry them open. I had the sense of wanting something to shatter, something else to come pouring forth.

———

AROUND A DECADE AGO, back when the Problem still felt new to me, I began paying attention to the sensation of heat. I knew this was only one of its symptoms—that the Problem was much more than a matter of degrees—but this seemed at the time like the simplest way to face its tangibility. On record-setting days, of which there were beginning to be more and more, I'd make a point of stepping outside for a walk, letting the sun strafe my head and weight my shoulders. I'd walk for blocks and blocks just to witness the temperature, the way it slackened my stride and faded through my soles.

When it's hot out, everything slows down: events unspool lazily, sounds seem to slur through a syrup. And yet heat is actually equivalent to speed—tredecillions of molecules vibrating faster and faster, the whole world buzzing at a higher frequency. I never knew how to square this. I just kept walking through the warmth, like I was passing through a series of heavy curtains, parting them with my face till I could no longer feel them on my skin. This too was a kind of numbness, though it was also a kind of crack.

On days like these hardly anyone else would be out, and I'd often have the sidewalk to myself, a whole baking lane of concrete. Still, when I did see someone coming toward me, I felt a greater than normal pressure to engage, make some gesture at the elephant in the room. Unbelievable, we'd say to each other, shaking our sweaty heads, though belief had long since ceased to be relevant.

It was extremely hot, in fact, on the day I started mapping my notes into this letter. I wrote all of my ideas down on little sticky notes and arranged them in various patterns on my wall, trying to impose some structure. But as soon as I'd stick one up, another would peel itself off, the heat unsealing it from the plaster, so that one by one the notes fluttered down into a jumble on the floor. Sorting through them, I couldn't remember what went where, or how anything related, and this produced a doubled despair. Because, it seemed, nothing I wrote would come closer to evoking the Problem than the pile of limp paper at my feet.

Toward the end of her life, the queer theorist Eve Kosofsky Sedgwick wrote an important essay on the difference between knowledge and realization: "The gap between knowing something—even knowing something to be true—and realizing it, taking it as real." It is the disregard for this gap, she argues, that marks one of the critical shortcomings of Western thought, the way we often mistake truth for reality. But they are two very different things: truth is a strictly epistemological affair, like a switch flipped on in the head—binary, cerebral, instantaneous. Realization, on the other hand, is a process, something that can only be achieved through practice, often over long periods of time.

It is possible to know something for months or eons and never realize it. Right at the beginning of a meditation retreat, for example, you are simply told a truth: about the impermanence of all things, the illusion of the self. But hearing this—even going so far as to "believe" it—has almost nothing to do with your ability to take it as real. Per Sedgwick: "To practice Buddhism, after all, is to spend all the time you can in the attempt to realize a set of understandings most of whose propositional contents are familiar to you from the beginning of your practice."

Months before writing the essay, Sedgwick herself had been diagnosed with terminal breast cancer. It was the approach of death that brought her closest to this crucial gap. Death, she wrote, "makes inescapably vivid in repeated mental shuttle passes the considerable distance between *knowing* that one will die and *realizing* it."

Can't the same now be said of the Problem, the most comprehensive form of death we've yet to invent? I don't mean this in the ordinary sense with which we exhort the denialists: that they need to "wake up" and recognize the truth of the Problem. The whole point is that, for most of us, the Problem is *merely* true. We've trapped it in a propositional form, caged it behind the bars of graphs. But you can examine all the graphs in the world and still never fully *realize* the Problem, at least insofar as you'd realize the sound of a fire alarm, or the tremor of an earthquake. This is a chasm far wider than the one between knowledge and ignorance, though we talk about it less often—maybe because it's too obvious, maybe because it's harder to bridge. Instead we keep writing the word "heat," as if this could ever be the same as feeling its referent wilt the paper.

There is a deep discomfort, I think, in trying to exist within Sedgwick's gap. To feel yourself so palpably alive yet so prosaically doomed. To look past the profusion of experience—its constant kaleidoscope of mood and color—and realize that what comes after it is nothing, not even blackness, just the absence of the absence of light. The temptation is to somehow resolve this dissonance, and there are many who, to this end, simply avoid the fact of death, retracting into life so as not to contend with its transience. But the other move is also possible, and I think proffers the more seductive numbness. To center death and treat life as the afterthought, like flipping through a book to its ending.

There have been times when I've succumbed to this numbness, when the certainty of death seemed like the best and perhaps only shell with which to contain the magnitude of the Problem. I remember that in my last year of college I read an essay called "Learning to Die in the Anthropocene." Its author, an Iraq war veteran named Roy Scranton, described how driving into Baghdad after the invasion "felt like driving into the future." The city "rose from the desert like a vision of hell: Flames licked the bruised sky from the tops of refinery towers, cyclopean monuments bulged and leaned against the horizon, broken overpasses swooped and fell over ruined suburbs, bombed factories, and narrow ancient streets." All told, Scranton served in Iraq for four years. In order to deal with the very real possibility that he'd be killed, he would meditate every morning on the inevitability of his death, as if to close Sedgwick's gap by force of will. "I'd imagine getting blown up by an I.E.D., shot by a sniper, burned to death, run over by a tank, torn apart by dogs, captured and beheaded, and succumbing to dysentery. Then, before we rolled out through the gate, I'd tell myself that I didn't need to worry, because I was already dead."

Two years after he returned—still alive—from his tour of duty, Hurricane Katrina hit New Orleans, and his airborne division was deployed to help prevent the outbreak of riots. It was like a flashback to the invasion, he said: "This time it was the weather that brought 'shock and awe,' but I saw the same chaos and urban collapse I'd seen in Baghdad, the same failure of planning and the same tide of anarchy." At the hands of the Problem, Scranton argued, the future would look just like this, just one long aftermath, an inevitable descent into social unrest and institutional collapse. Organizing was therefore a waste of time, mourning being the only real option left to us. It was already too late for any technological breakthrough, any long-shot social uprising. We needed to realize, he said, that "this civilization is already dead," that "there's nothing we can do to save ourselves."

"Learning to Die in the Anthropocene" was published as an essay in 2013, not long after Sandy had wrecked the eastern seaboard and lent credibility to Scranton's vision. When I read the piece, it was the first time I'd seen anything like it set down in print. The raw pessimism felt scandalous, like Scranton was saying the unsayable, like he'd cracked open a Pandora's box. I remember having to close my laptop and take a walk around the block, hoping that if I came back and read it again the essay would somehow cancel itself out, allowing me to scrub its conclusions from my memory.

Instead they stuck with me. For years the Scranton essay represented one pole on the spectrum of attitudes I could adopt toward the Problem. It was here I was turning in my moments of despair: when I flirted with anti-natalism or pictured killing myself like David Buckel, when collapse seemed so assured as to already have happened. This numbness came to a head with Trump's election. In the months after his inauguration, the phrase "already dead" kept repeating itself in my mind, echoing into a mantra. This was one way to move through the hopelessness: *I'm already dead*, I'd recite to myself as I logged into my conference calls; *I'm already dead* as I explained to yet another state legislator, in a tone that fell short of full conviction, that now more than ever New York had to lead in the fight against the Problem.

For a time, it really did help. Being dead lowered the stakes on everything, softened out all the edges. Whenever I felt close to panic, I could switch instead into a posture of wistful regret, like I was looking back at the world from beyond the grave, shaking my head at its mortal foibles. For months I floated around in this feeling, which I recognize now as a species of grief, although at the time it felt more like weightlessness. I continued doing all my work for NY Renews, never missed an email, but it was like I wasn't really there, like I was some sort of ghost stalking the edges of the Problem, a haunting of the haunting.

This attitude was impossible to sustain for very long. Eventually, inevitably, I sank back into my body, which was indeed still alive and had been going quietly about the business of feeding and resting itself, even as my mind prevaricated. I allowed myself to be consumed again by the work, to feel the stakes at my back and let them thrust me forward.

It wasn't that I no longer believed Scranton's basic assessment of the Problem. It was just that, with time, I'd begun to feel the difference between acknowledging death and *surrendering* to it. There was something in Scranton's response that felt too resigned to me, too bereft of agency. When I finally rejected it, the experience was almost involuntary, like I was coughing up something I'd swallowed.

Because of course the end would come—in some form, at some time—but wasn't the whole point to postpone it? Wasn't there something essential, some foundational spark of the human experience that sprang from exactly this friction, between the fact of death and the will to live? I was not, I decided, going to let the Pruitts steal even this central tension. They'd already stolen the coasts, and the seasons, and the lives of the drowned, but this at least was something those fuckers would never get. I was going to keep rubbing death against life until they burned with each other, until they fused. Who was I—and, for that matter, who was Scranton—to let these sparks go out, when there were millions still maintaining them in the very jaws of the Problem, on land that'd been parched, burned, or flooded? These people were not already dead; they were alive—insistently alive—and if they hadn't given in, then certainly neither could we.

I AM WONDERING WHY I FEEL the need to zoom out like this, to re-trace all these steps. I suppose I wanted to offer you some parting advice, as I know you're meant to in this sort of letter.

Most letters addressed to children are written by someone with a lifetime of lessons to impart—a wise old parent, maybe, or a retiring mentor. I am neither, obviously. But the Problem is speeding everything up, and so I've felt the need to write to you now, in my twenties, though I'm still circling the Problem my-self, processing down a path whose end I may never see.

In case it's of use, then, here is one piece of advice: Do not accept the vision of our future as a single road leading to a burn-ing city. Compromised as it is, it still seems to me more like a fan, stretching out in front of us in a swath of possible outcomes—most of them scary, maybe, but none of them entirely predict-able. In this indeterminacy, there is potential, which means there is still room for movement. Do not feel compelled to suffer Scran-ton's passivity on this point. However grave the dangers become, however dire the warnings, they will never collapse your future into something as narrow as a fate.

Still, there will be days when they'll seem to, and this feels important to name. I'm remembering the morning after that crushing election, how my own future really did feel flattened.

Unwilling to go to work and unable to be alone, I called my sister at her apartment in Brooklyn. Paradoxically, her baseline anxiety seemed to lend her remarkable powers of calm whenever the facts of the world conspired to vindicate it. I could feel my own anxiety flailing around, frantic, trying to press the world into some alternative configuration. You should come over, she said. She sounded soothing and resourceful, like Kirsten Dunst's character in the final scene of *Melancholia*. I told her I loved her and caught the subway down from the Bronx.

That day there was a new kind of silence on the trains. More piercing, somehow, more frightening. People fixed their stares and didn't let go. A woman cried quietly in a corner seat and no one reacted. It was like we were paralyzed, like a single false move would crack us all open.

When I got to her apartment my sister was sitting for a family with a baby named Joan. For hours I lay on the living-room carpet, baby Joan crawling over and around me, fitting various plastic toys into her mouth. In my body the numbness was registering as a weight, like an anchor holding me to the floor. I thought about moving but didn't know how or why I would. So I stayed splayed, feeling dead and knowing I wasn't and wishing I was.

Eventually, the afternoon rolled around and we mustered the energy to take the baby out in her stroller. It was raining lightly that day, and my sister led us the few blocks over to Prospect Park. We walked in loops around a duck pond fringed by willows, then wandered off into the maze of empty trails where, months later, David Buckel would take his own life. Baby Joan kept pointing things out from under the bonnet of the stroller, thrusting her stubby finger at a squirrel or a trash can. My sister managed a smile every time, bending down and pointing with her at whatever had caught her interest. Under the circumstances, this struck me as a near heroic feat of cheerfulness. I, on the other hand, was completely useless, already drifting out of myself, my whole attention focused on keeping the stroller on the path.

We spent the rest of the afternoon this way, walking dazed through the rain, treading lightly without intent. The sun passed in and out of the clouds. One by one, Baby Joan singled out the things that hadn't changed: the benches, the lampposts, the wet slicks of grass—everything wholly oblivious, everything obliviously whole. The future shuddered, but seemed to hold.

———

IN THE MONTHS FOLLOWING THAT DAY, a new tone arose among the ensuing marches, mass emails, and rush-to-print resistance guides. Everyone started urging everyone else to stand up and fight back, to get angry and active. "Don't mourn, organize!" we repeated to ourselves, to one another. This was the exact opposite of Scranton, like we'd all agreed to slam shut his Pandora's box, to secret it away and never look inside. So whenever we spoke— as if in compliance with some implicit rule—we tacked on inflections of hope to the ends of our sentences, such that even our sincerest expressions of optimism were soon inflated into a tender of little worth, pennies of faith we exchanged mostly as a matter of form.

I understood this impulse, felt it deeply in myself. It was hard enough to take what was happening as true, let alone to take it as *real*. But I also knew, in the back of my head, that it wouldn't sustain us. If Scranton was surrendering to the possibility of death, then we were simply ignoring it, rushing to the barricades without allowing ourselves to imagine what lay behind them. The kind of strength this took—and it undoubtedly did take strength—felt somehow brittle, almost too hard, like it could crumble at any moment. And when we raised our fists in defiance, when we shouted out our chants, I wasn't always convinced that what we were resisting was the Problem itself. Perhaps by then we'd realized that the grief was another kind of tide, that it too could be dammed.

These, then, were the two choices on offer: you could mourn or you could organize, you could weep or you could work, you could admit the full, crushing truth of the Problem we were actually facing—let it cave in your walls and drown your abjurations, so that even in your own head there would be nowhere left to run— or you could dam it up behind something labeled hope or courage or determination so that you would never have to look at the thing itself, just the shell you'd erected around it, which like any flood wall betrayed nothing of the currents at its back.

I was caught between these options for years, too hopeful to cave, too hopeless to simply put on a brave face. I wanted to mourn *and* organize but could do neither fully, mired as I was in the sense that they precluded each other.

Fuck this choice. I reject it, and I want desperately to keep you from falling into its trap. It is a terrible, impossible dichotomy— and what's more, it is false. I no longer believe that grief and resistance are mutually exclusive: I think the former is *necessary* to the latter, that honest sorrow is perhaps *the only thing* that makes a real fight even possible. To mourn without fighting is to tap out at the exact moment we need to step in, but to fight without mourning is to grapple with a ghost, to try to stop something you've never actually realized. Because how can you solve a Problem you do not take as real? And how can you take the Problem as real without feeling like a dam has finally broken inside you, like a flood is overwhelming that city you call your self? The way it drowns all your plans and presumptions, inundates them until they can no longer be recognized, until they're just strange shapes sticking up above water. The way afterward it leaves everything dripping and washed, somehow both brand-new and completely destroyed.

This is a terrifying prospect, but I'm not convinced there's another way: to contend seriously with the Problem you'll first have to let it in. And when I say let it in, what I really mean is drag it toward you, press it down, sit with it past the point of discomfort and pain and despair until you can observe it without blinking, until its weight is just a thing about you. In this way, "letting something in" is too passive; what I'm talking about is fitting a hyperobject into your heart without it breaking.

I realize that there is something counterintuitive about this particular token of advice: to drink grief itself as a nepenthe, to soften up just when the prevailing wisdom is telling you to steel yourself against the coming storms. In truth, I hope you'll never steel yourself. Steeling is what the Pruitts do, simply clenching their jaws against fact and feeling. The gulf between knowledge and realization is vast, and it can take ages to close, but you should try instead to be gentle with yourself. I am saying this as much to myself now as to you. Don't let your contempt for the dissonance banish it beyond your curiosity or resolve it easily into thin air. As Sedgwick wisely suggests, "Perhaps the most change can happen when that contempt changes to respect, a respect for the very ordinariness of the opacities between knowing and realizing."

Recently I read an article on the psychologist Renée Lertzman, an expert in what she's termed "environmental melancholia." She offers the following advice: When you're trying to have an honest conversation about the Problem, it helps to first acknowledge the weight, to let the dam break a little. Starting off with something between an admission and a commiseration, she suggests—something along the lines of "damn, this is intense"— "frees up a lot of energy to move into problem-solving mode."

This is sage advice, I think, and it's more or less what I've been trying to do: to create a shared foundation of vulnerability from which it might be possible for us to discuss the Problem not simply as a "current event"—that mollifying phrase—but as something approaching a real experience.

But I'm also wary of instrumentalizing things too quickly, the eagerness with which grief becomes something to harness, something that can be put to work—as if fear and sadness were just another resource to be extracted and burned, used, as Lertzman says, to produce "a lot of energy." I've told you how, in the wake of Hurricane Maria, I watched a friend elegize his grandmother to a governor who wasn't listening, in hopes that this would someday allay the kind of storm that had killed her. I've watched myself—over the course of many nights spent hunched in front of my laptop—translate grief into grind and back again. And I want to tell you that this is not required, that the validity of your feelings is not reducible to their utility in the movement. Fear and sadness can be commendable tools, but they are also things unto themselves—stones, or shells, or clods of earth, things with weight but no telos. You do not have to mold them into anything; you do not have to use them to build. If you want, you can just let them sit there—an inner geology—accreting into lines that you will one day be able to read.

In the process of writing to you, I've sometimes felt myself wanting to communicate up a generation as well, so recently I've been trying to be more open with my parents. If what I'm feeling when they call is grief, this is what I tell them, right off the bat, per Lertzman. It never seems to translate over the phone: I usually sound more clinical than I feel, merely relaying a fact.

Still they are sympathetic, and try to engage. They tell me that I shouldn't punish myself, shouldn't feel like it's all on my shoulders. How can we cheer you up, they always ask. My dad suggests a run, my mom suggests something on Netflix. Anything to take my mind off the Problem. Because they hate to see me grieve like this, they tell me; they're not sure whether it's productive.

I disagree with them, but I understand. They are worried. They don't want me to wallow. This is an expression of their love: that when I try to talk to them about the Problem, all they ever end up talking about is me.

By now, I've also told them about the times when I've doubted I should have you. And though they never say it outright, I know they want you to happen, want to be able to hold you in their arms someday. Perhaps they're worried that if I'm ever allowed to sink too far into realization, then this possibility will disappear. What they don't understand is that the opposite is true. That if the numbness remains intact, then you'll stay trapped in it forever. If you're born at all, it will only ever be through a crack, a pouring forth.

Rarely do I share these anxieties with any boomers beyond my parents. Their generation did not grow up conscious of the Problem, and will not live to watch it metastasize. Usually, people over fifty just tell me not to worry about having kids. "It's *their* generation that's going to solve the Problem!" they'll say, smiling in a way that suggests I lighten up.

I hate this attitude—how breezily it passes the buck; how it lets itself off the hook, swallows optimism like a sedative. This shouldn't be how it works, each generation abandoning the next to an increasingly impossible situation, waving goodbye and good luck. True intergenerational justice demands more of us. It can't just be the long line, the human string spitting itself further and further into the future. There are other shapes a legacy can take. A widening gyre, maybe, a Mobius strip. I don't know, exactly, but I feel like I can catch the edges of them, like the Problem is beginning to make them thinkable.

Whatever form it takes, my own legacy can't end with this letter. Once it's finished, I'll put my pen down and get back to organizing. The world from which I retreated is the same to which I'll return, and I can't live in it for long without working to prevent its unwinding.

Fortunately, as of this writing, the movement continues to grow. Many millions of people are refusing to give in, and I plan to go on being one of them. This feels especially crucial if I'm to have you. What kind of dad would I be, letting you do all the work?

———

SOMETIMES I CONFIDE IN PEOPLE my own age, though even we have our substitutes for mourning. Increasingly we talk about rage. Rage at the rich men who keep drilling no matter what, out of greed or spite or mulish inertia; rage that they would sacrifice millions of lives for another win, another yacht, another desperate postponement of the day when the stories they've been telling themselves will implode.

I have nothing against rage. Rage motivates, it gets things done—and my god is it justified. But in the long run, it cannot be the primary basis for our movement. I am not making a moral argument here, merely a psychic one: that rage is unsustainable, that it cannot guide our work through the coming centuries. The very objects of our rage won't last. The Pruitts will lose, and this is a question of when, not if. In the most prosaic and obvious sense, failure is simply baked into their plans. Every interim victory they score—every pipeline they build, every industry they deregulate, every wilderness they rip open for mining—brings their system closer to the brink of ruin. This cannot but be the case. They've premised the whole enterprise on a patchwork of absurdities: the possibility of infinite growth on a finite planet; the denial of collectivity in an interdependent world; the relegation of that large portion of experience that cannot be shoehorned into monetary valuation—the things the rest of us might call beauty, or strangeness, or joy—relegating these to the status of "externalities," essentially ignorable.

The economy they've constructed from these premises—their

greatest triumph, the thing they defend so frantically with their money and their propaganda, the thing for which they claim there is "no alternative"—is eating itself. And in an autophagic system, every win is a loss, every bulwark a wrecking ball.

This is not to say we should be complacent: there is still the question of how soon they will fail, and how many they'll bring down with them in their nihilistic death spiral. But it does mean that at some point the Pruitts will fall away, and if all we have is rage, then we'll be left with nothing, no sense of how we should move forward, or why. If we are going to outlast them, then our emotional sustenance cannot depend on them.

We'll need something more durable, something that will see us well past the shortening lifespan of the Pruitts, into a future we can't yet see. Which is why I've arrived here, at that most predictable and deeply felt cliché: that it is more powerful to love the things you might lose than to loathe whoever's trying to take them.

And so now I am clear on what I love, I have to be.

I love you. I love you even though I don't know you, even though you're still just a choice. What I think I mean by this is that the thought of you makes me less important to myself, turns down the volume on that incessant engine of subjectivity—the weaver of narrative, the hoarder of self—so that for a moment I can tune out its roar and inhabit the wonderful, dissolving quiet of the space beyond my head, where relation splits open identity and grace comes flooding through the breach.

Though it should be said that what I'm in love with is not you, really, but the idea of you. And falling in love with the idea of you is just the prologue. When and if you are born, the actual you will break through the hypothetical you in ways I'll never be able to control or anticipate, and this to me is one of the chief appeals of parenting.

So take this letter however you will: reject, subvert, rebuild, extend, or ignore it. I want to avoid this being one of those cases where the paternal image of the child ends up stifling the real thing. Like that phrase you'll sometimes hear parents say: "I see myself in you." I've never liked this expression—what I want to see in you is you. A person I can try to imagine but will never presume to define.

Because what is imagination except a conjuring of the hypothetical, a line of attention spooled out beyond the senses? And conjuring you—my own, hypothetical child—is only one small step in a much larger task of imagination. To live in the Anthropocene is to realize that your attention must be broadened far beyond the bounds of your individual circumstance—expanded to encompass people, species, objects, and eras with which you are both utterly unfamiliar and inextricably bound. The hares in the mountains. The fossils in the cliffs. The people of Far Rockaway, and Australia, and the future, and the Now. This practice is both daunting and divine, quotidian and unending. In this, it resembles religion. As the philosopher Simone Weil wrote: "Attention, taken to its highest degree, is the same thing as prayer. It presupposes faith and love."

Here's an irony I've noticed about this sort of attention: imagining you is one of the things that make the Problem most *real* to me—that jolts me awake to the sheer immensity of its stakes—and yet it is only in the shadow of this immensity that I've ever questioned whether I should have you. I'll admit I don't always know what to do with this, how the idea of you contains both rationale and refutation.

I do know that I've been afforded considerable help in my attempts to transition from knowledge to realization. Storms and marches, victories and defeats, certain scenes in movies and movements in music and moments I've spent alone outdoors—all of this, at one time or another, has helped me narrow Sedgwick's gap. But there is something about you in particular that makes the weight of the Problem palpable, like it's moved from my brain down into my bones.

Whatever it is, I want to thank you for it. Writing to you has helped me live in reality, to behold the Problem without blinking or turning away. In this there is a second irony: that for all my talk of bringing you into the world, it is you, ultimately, who's helped bring me into mine.

———

WHEN I IMAGINE YOU NOW, you are not a long line or a next step; not a martyr or a savior or an emissary. I picture you instead as the glint of something larger, a pattern just emerging into view. It is all of us—the living and the dead and the not-yet-alive—all of us rendering one another whole, backward and forward through time, no teleology, no direction even, just a gradual process of collective realization in which birth is just a single, glancing stage; in which not just responsibility but identity itself floods out past the walls of longevity—way, way out, to where the waters of memory and foresight converge, to where we're all just one flat surface above a fathomless depth and we're rising, always rising.

Past this horizon I'm not sure we can see. There are certain things that the Problem will keep hidden, endings it won't yet divulge. But this much we know: One day we will all be gone, and there won't be any gap between what is real and what is realized. Consciousness, that old funhouse mirror, will no longer stand in the way. Things will just be as they are—finally, effortlessly; they will admit no additional meaning.

And if by then the sky has rejected our input, and the oceans have resorbed our mistakes, and the crust has buried the carbon in whose black ink we wrote our sentence—then and only then might the seasons return. I imagine they will come back slowly, laying down their old moods into fields of shrub and stone. The last things won't notice; they won't do anything at all.

Leaves will fall to no rakes. Wind will whip at no flags. Rain will runnel no gutters. And once again the snow will emerge to the foreground, loosed of its hunters, a witness-less white blinding no one, shrouding nothing. The flakes will pile and freeze and there won't be anyone there to paint them. They will just accumulate, layer after layer, the softened outlines of a once-sharp world.

Until then, there are lifetimes to live—maybe many, maybe yours. Until then there are families, still, and first birthdays, and great banks of cloud. Beneath them the birds will take shelter, and the signal will cut out, and the planes will be too hot to fly.

And it's true that I cannot conceive what I cannot conceive, but even through the Problem I think I can still see you, hazy as a mirage, afloat in the world to come. So if I began this letter to show you where you've come from, perhaps now I can venture at where you'll go.

How you'll still keep a jacket for winter. How you'll no longer name all the storms. How you'll watch Latour's vortex collapsing, watch geology crash through history into biography, and how as this happens, as time finally drops its scales, you will feel yourself alive in its confluent stream—born of it, borne by it—perhaps for the first time.

And soon the earth's pace will draw near to your own, and there will be no distinguishing the two. Rivers will run their banks and mountains will slide down themselves and baby teeth will wiggle loose and fires will leap from their rings and rings will be fitted to fingers and the smoke will ruin the photos and the glaciers will wind up their gullies and the videos will play on fast-forward and the cat will lie down on the carpet and the ocean will rise from its knees and you will wake up exhausted, listening for rain.

Maybe then, hearing nothing, you will take out this letter and pore over it yourself. I don't expect it will hold. You'll stream right through it, page after page—each one a flimsy dam, a Now sliced thin as paper—until one by one they'll all falter and breach and you will flood out past the final words into that blank gulf called the future, or the possible, or the end.

And this is not where I leave you. This is where we meet:

Acknowledgments

Thank you to the Adnyamathanha, Kaurna, Eansketambawg, and Coast Miwok peoples, on whose unceded lands I wrote this book.

To my agent, Veronica Goldstein, whose savvy, candor, and almost frightening diligence helped guide it to publication through a year of profound upheaval.

To the entire team at Penguin: Louise Braverman, Beth Caspar, Kasey Feather, Britta Galanis, Bridget Gilleran, Brianna Harden, Daniel Lagin, Randee Marullo, Patrick Nolan, Lindsay Prevette, Kate Stark, Mary Stone, Brian Tart, and, of course, my inimitable editor, Allie Merola. On page after page, your curiosity and thoughtfulness made this book better. And our meetings—on my roof, in our deck chairs, over bags of Doritos—were the highlight of the publishing process.

To every teacher who's ever given me a novel or a writing prompt. None of you are paid enough for your work.

To Brown University, which at long last divested from fossil fuels. And to every cultural, educational, political, and financial institution that still hasn't: What are you waiting for?

To Kurt Ostrow, who believed in this book long before I did, and then dragooned me into writing it. Thank you for always picking up the phone, for workshopping the tiniest details, for hosting an unhealthy number of book clubs, for engaging with my Goodreads reviews, for never forgetting snacks, for practicing unabashed gratitude, and for being a tireless, radical, and beloved union educator. You are my best friend, and quite possibly *the* best friend. I love you and feel so lucky to have you.

To Emily Kirkland, who saw what this book was trying to do, and gave critical feedback on an early draft. The Problem would feel a lot harder if we couldn't laugh/cry/excoriate Mitch McConnell together. I always feel less alone after I call you.

To Becca Rast and Maya Sikand, who introduced me to organizing and shaped my concept of politics. To the rest of my chosen family: Keally Cieslik, Emily Oglesby, Gabe Schwartz, and Luke Taylor. Youse all keep me grounded and afloat.

To Jordan Breslauer and Scott Shapses, who made Jersey tolerable.

To Megan Hauptman, who taught me a conscientiousness I'm still working to live up to.

To Cimarron Forbes, who's shown me the meaning of resilience.

To all the friends and mentors who gave feedback on the manuscript or supported me in writing it: Sam Adler-Bell, Graham Akhurst, Alda Balthrop-Lewis, Camila Bustos, Tyrese Coleman, William Dinneen, Cory Hargus, Timothy Herbert, Catherine Imbriglio, Peter Kentros, Myles Lennon, Sergio López, Kim Mahood, Giovana Schluter Nunes, Jenny Offill, Rachel Schragis, and Alexis Wright.

To the Aussie mates: Reece Kinnane, Claire Bowman, Rachel Bala, Rana Kokcinar, Josh Baldwin, Colline Bertin, Benjamin Madden, Kat Beazley, and Lauren Schilds. See you at the Ex someday.

To Stephen Muecke and Prudence Black, who welcomed me in loco parentis.

To the Adelaide Hills, where I walked when I couldn't write.

To the Australian-American Fulbright Commission, the Mesa Refuge residency, and the J.M. Coetzee Centre for Creative Practice, which gave me the time and space to think things through.

To Daniel, Errol, Terry, Janice, Richard, the rest of the Roe and Hunter families, and the entire Goolarabooloo mob. Thank you for taking me out on country, and for helping me hear its voice.

To Dr. Rosiana Lagi and everyone at the USP Tuvalu Campus, who model hope, humor, and grit on the very front lines of the crisis.

To all the organizers, artists, strategists, wonks, trainers, legislators, elders, and young people in the movement. Thank you for working so fucking hard, against the clock, to build a world where everyone can thrive. Your courage and compassion astound me. You embody the society I want to live in. Thank you in particular to the amazing people—too numerous to name—who I've worked with at NY Renews, the Green New Deal Network, the Sierra Club, South Bronx Unite, and Brown Divest Coal. And to Daniela Lapidous, who stepped up when I stepped back, and who helped get the Climate Leadership and Community Protection Act across the finish line.

To Babbie and Maury, who anchor the mishpacha.

To Mom and Dad, without whose love and support this book would never have been written. You taught me how to ask questions, then gave me room to explore. You taught me to live by my values.

To my sister, Izz, who I've been looking to since the beginning, and who always knows what to say.

And to Mengi, my partner, whose love helps realize the world.

The climate movement is fighting for the future
of human civilization, and we need your help.

Whether you can give time or money,
here are two good places to start:

www.sunrisemovement.org

www.climatejusticealliance.org